AIRCRAFT INSTRUMENTS

AIRCRAFT INSTRUMENTS

BY

C. A. WILLIAMS

A.F.R.Ae.S., M.S.L.A.E., A.M.Brit.I.R.E.

ODHAMS PRESS LIMITED

LONG ACRE, LONDON

First published 1961

T.661.N.

MADE AND PRINTED IN GREAT BRITAIN BY
THE GARDEN CITY PRESS LIMITED
LETCHWORTH, HERTFORDSHIRE

Contents

Foreword

THIS book is devoted to an explanation of the principles of general-purpose instruments used in aircraft to-day, and of the rules and procedures governing the maintenance of those instruments. As such, it is intended for use by airline instrument engineers, by those who wish to become instrument engineers, and by all who are interested in, or concerned with, general instrument practice. Assimilation of the information contained in its pages should lead to the gaining of the maintenance engineers' Licence in category "X" (instruments—including those electrically operated).

The book is not intended to do more than establish or confirm principles so far as design engineers are concerned, nor is it a platform for discussion of the relative merits or demerits of the various instrument manufacturers' designs and products. If, in the course of the book, it is noticed that one manufacturer's designs are illustrated rather more often than those of another, it is simply because I have had more ready access to those illustrations.

I must here place on record appreciation of the help received from manufacturers in supplying both illustrations and technical literature. Accordingly my sincere thanks go to:

Air Registration Board.
British Standards Institution.
Bryans Aeroquipment Ltd.
De Havilland Aircraft Co. Ltd.
Kelvin & Hughes (Aviation) Ltd.
Negretti & Zambra Ltd.
Sangamo Weston Ltd.
Simmonds Aerocessories Ltd.
Smiths Aircraft Instruments Ltd.
Vickers-Armstrong (Aircraft) Ltd.
Waymouth Gauge and Instrument Co.

In addition, I wish to express gratitude to my wife who "translated" my rough notes, to Miss Price who typed and retyped the manuscript, and to my colleagues Mr. Bob Jones who produced many of the illustrations from my rough sketches and Mr. Tom Sharpe who co-ordinated all these efforts and without whom the work could not have been finally prepared for publication.

Cheltenham, 1961. C. A. W.

Part I

MECHANICAL INSTRUMENTS

CHAPTER 1

Aircraft Instrumentation: Terminology

THE aircraft pilot of the twenties had what is by modern standards a primitive machine, with virtually no instruments at all. He was, in fact, said to fly "by the seat of his pants". To-day an airline pilot, his navigator, and his flight engineer, are faced by a bewildering display of flight instruments and engine instruments, together with radio, navigation aids, automatic pilots, flight systems and the like. The modern flight crew can therefore be said to fly "by the look on a dial".

It is certainly true that as aircraft complexity and traffic loading grows, the number, scope and complexity of instruments fitted to aircraft also grows. In many instances this is solely due to multiplication; for example, an aircraft with one engine needs only one oil-pressure gauge whereas an aircraft with four engines needs four gauges. Nevertheless, there is a school of thought which believes that the increase in instruments has been overdone and that a halt must be called before the pilot becomes a mere "meter inspector". This may be true to some extent, but it must also be remembered that the safety and economics of modern airline operation demand extremely close limits on navigation, engine performance and flight control, so the pilot must be provided with accurate, readily available information on all these factors with the minimum of delay. Again, since an aircraft is an asset only when it is in the air, the means of providing the information must be reliable and long-lived if they are to be economically worth while.

The modern aircraft instrument is the means of providing the information required by the pilot at any given moment. It is also designed to be reliable for a given period or "life" to avoid causing expensive operational delays through instrument failure.

Range of Instruments. Some idea of the range of instruments carried by a modern aircraft may be obtained by studying the illustrations on the following pages of the main instrument panels in the De Havilland Comet aircraft.

It will be seen from the list of items on the pilot's main panel that there are some 27 indicating lamps for various warning functions. These may be operated simply by a micro-switch in the case of "landing-gear up" indication, pressure-operated switches in the case of low oil pressure, or radio transmission in the case of the I.L.S. marker and beacon identification lamps. Indicators of the type shown at (20) I.L.S. deviation indicator, (24) Search radar indicator and (29) Decca flight log, are not usually the concern of the instrument mechanic but rather of the radio specialist. No mention of these instruments will therefore be made in the following chapters. Other instruments such as the R.M.I., O.B.S., and R.M.S., are also concerned with radio navigation but brief mention will be made of these instruments as an indication of the use which is made of synchronous systems.

Certain other instruments such as the director horizon, beam compass, S.F.S. annunciator, and engage and trim indicator, are part of an integrated flight system and auto-pilot. Such equipment is to be described in a companion volume to the present book,* now in course of preparation.

In the case of the fuel panel, it will be noticed that many of the indicators are again of the warning lamp or magnetic flag (doll's eye) type. The main instruments comprising fuel flowmeters, thermometers, and contents gauges are described later in the text under appropriate headings. The electrical panel is mainly concerned with the instrumentation which indicates the state of the electrical supplies throughout the aircraft. Most of the instruments on this panel are therefore the concern of the electrical technician, but it will be seen that there are generator thermometers and hydraulic pressure gauges which come within the sphere of the instrument mechanic. Again, in the case of the pressure, temperature and de-icing panel many of the indicators are of the magnetic flag or doll's eye type, but the thermometers, cabin height indicator, rate of climb indicators, etc., are described later in the text under appropriate headings.

Medium-range Airliner. In the case of a medium-range airliner, such as the popular Vickers Viscount, the engineer's panels are integral with the pilot's and co-pilot's panel as shown on page 19. The key drawing accompanying the photograph indicates practically the full range of instruments carried by this aircraft.

It is hoped that these illustrations and details of the instruments carried by the Comet and Viscount will prove of interest as giving an idea of the number and types of instruments to be found on modern aircraft, but it should not be assumed that the instruments described in the text are necessarily the same

* *Aircraft Instrument Control Systems*, by C. A. Williams. To be published in 1962.

as those fitted to either of these aircraft. The treatment is more general than this so that the reader may apply its principles and practical hints to the particular aircraft with which he is associated.

MANDATORY INSTRUMENTS

Not all instruments carried by the modern airliner are "mandatory" instruments, i.e. instruments required to be carried by current regulations. Information regarding these mandatory instruments can be obtained from the appropriate Air Navigation Orders (No. 829, 1954) and Air Navigation Regulations (No. 925, 1954). Additionally, reference should be made to British Civil Airworthiness Requirements (B.C.A.R.) and the following references in that manual are of interest to the "X" licensed engineer:

(i) *Chapter A6-5.* Specifications (Instruments, Equipment and Accessories).

(ii) *Chapter A6-6.* Specifications (Materials).

(iii) *Chapter A7-3.* Design Organizations (Instruments, Equipment and Accessories).

(iv) *Sub-section D6.* Equipment Installations. Chapter D6-1. General. (Paragraphs 1.1 to 6.4.)

This last chapter is concerned with Non-mandatory and Mandatory Equipment.

Para: *2. Non-mandatory Equipment*

Where non-mandatory equipment is installed it shall be demonstrated that the equipment as installed is neither a source of danger in itself, nor a prejudice to the proper functioning of any Essential Service, and does not in any way reduce the airworthiness of the aeroplane to which it is fitted even in the event of its failure to function.

Para: *3. Mandatory Equipment*

3.1. Mandatory Equipment is defined as the minimum equipment necessary before a Certificate of Airworthiness can be issued, as prescribed in 4, 5 and 7 (of Chapter D6-1 in B.C.A.R.).

3.2. All mandatory equipment shall be of an approved type. It shall be established that "installations incorporating Mandatory Equipment and equipment installed in the aeroplane to enable the fitment or use of other items prescribed by Air Navigation Regulations" will adequately and reliably perform the functions prescribed.

Other paragraphs in Chapter D6-1 of B.C.A.R. are concerned with pre-installation inspection, instrument locations, instrument illumination, panel vibration, instrument power supplies, duplication of instruments and air-speed indicator and altimeter errors. Paragraphs 4, 5 and 6 of this Chapter are concerned with the minimum equipment to be carried on aeroplanes.

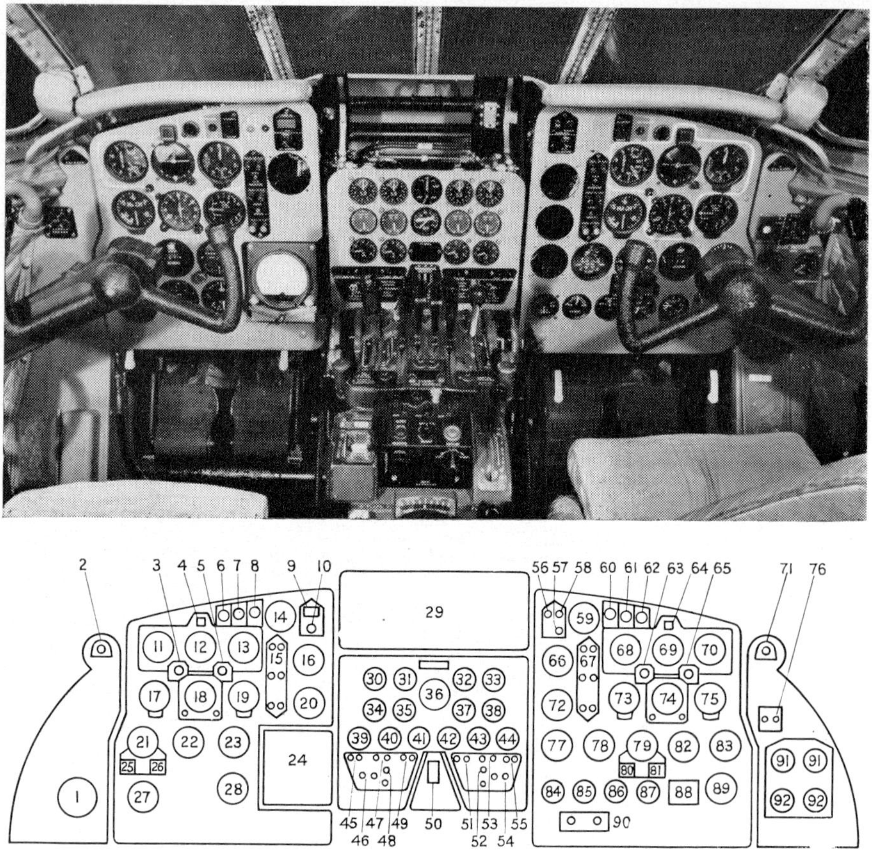

DE HAVILLAND COMET MK. 4

Pilot's Main Instrument Panels

1. Turn-and-slip indicator.
2. Stall warning lamp.
3. "Landing-gear up" warning lamp.
4. VOR/ILS coupling indicator.
5. "Aircraft below 10,000 ft." warning.
6. "Blue" hydraulic system low-pressure warning lamp.
7. "Green" hydraulic system low-pressure warning lamp.
8. A.C. supply failure warning lamp.
9. "Check elevator gear" warning lamp.
10. Ice formation warning lamp.
11. Airspeed indicator.
12. Director horizon.
13. Altimeter.
14. Elapsed time clock.
15. I.L.S. marker and beacon identification lamps.
16. Radio altimeter.
17. Machmeter.
18. Beam compass.
19. Rate-of-climb indicator.
20. I.L.S. deviation indicator.
21. VOR/ADF radio magnetic indicator.
22. Omni-bearing selector (O.B.S.).
23. Stand-by artificial horizon.
24. Search radar indicator.

Pilot's Main Instrument Panels (*Contd.*)

25. "A.D.F. selected" indicator.
26. "V.O.R. selected" indicator.
27. S.F.S. annunciator unit.
28. Engage-and-trim indicator.
29. Decca flight log.
30. No. 1 engine r.p.m. indicator.
31. No. 2 engine r.p.m. indicator.
32. No. 3 engine r.p.m. indicator.
33. No. 4 engine r.p.m. indicator.
34. No. 1 engine power loss indicator.
35. No. 2 engine power loss indicator.
36. Engine synchronizer.
37. No. 3 engine power loss indicator.
38. No. 4 engine power loss indicator.
39. No. 1 engine exhaust thermometer.
40. No. 2 engine exhaust thermometer.
41. S.F.S. selector switch.
42. Air temperature thermometer (atmospheric).
43. No. 3 engine exhaust thermometer.
44. No. 4 engine exhaust thermometer.
45. No. 1 engine overheat and turbine failure warning lamps.
46. No. 1 and No. 2 engine top temperature control switches.
47. No. 1 and No. 2 engine oil low-pressure warning lamps.
48. No. 1 engine reverse thrust position indicating lamps.
49. No. 2 engine overheat and turbine failure warning lamps.
50. Landing gear mechanical position indicator.
51. No. 3 engine overheat and turbine failure warning lamps.
52. No. 4 engine reverse thrust position indicating lamps.
53. No. 3 and No. 4 engine oil low-pressure warning lamps.
54. No. 3 and No. 4 engine top temperature control switches.
55. No. 4 engine overheat and turbine failure warning lamps.
56. "Cabin de-pressurized" warning.
57. Ice formation warning lamp.
58. "Pressure doors open" warning.
59. Elapsed time clock.
60. A.C. supply failure warning.
61. "Green" hydraulic system low-pressure warning lamp.
62. "Blue" hydraulic system low-pressure warning lamp.
63. "Landing-gear up" warning lamp.
64. V.O.R./I.L.S. coupling indicator.
65. "Aircraft below 10,000 ft." warning.
66. Decometer (Red).
67. I.L.S. marker and beacon identification lamps.
68. Airspeed indicator.
69. Director horizon.
70. Altimeter.
71. Stall warning lamp.
72. Decometer (Green).
73. Machmeter.
74. Beam compass.
75. Rate-of-climb indicator.
76. "Red" hydraulic pump switch and indicator.
77. Decometer (Purple).
78. Flap indicator.
79. V.O.R./A.D.F. radio magnetic indicator (R.M.I.).
80. "A.D.F. selected" indicator.
81. "V.O.R. selected" indicator.
82. Omni-bearing selector (O.B.S.).
83. Turn-and-slip indicator.
84. Elevator gear-change indicator.
85. Yaw damper indicator.
86. Yaw damper switch unit.
87. Cabin height indicator.
88. Accelerometer.
89. S.F.S. annunciator unit.
90. Relight over-ride switches.
91. Wheel brakes—"Green" hydraulic system pressure gauges.
92. Wheel brakes—"Red" hydraulic system pressure gauges.

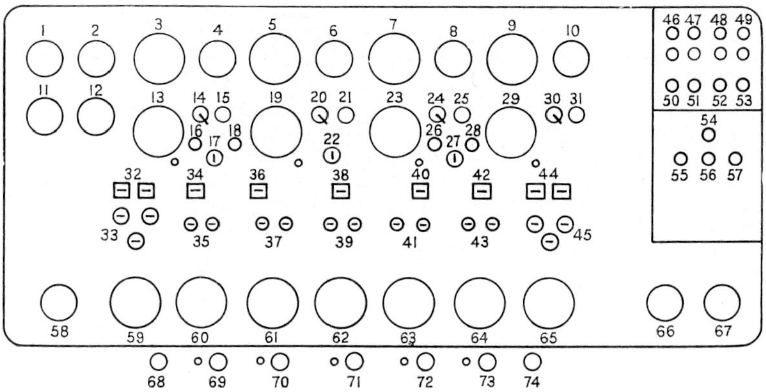

DE HAVILLAND COMET MK. 4

Fuel Panel

1. "Yellow" hydraulic system contents gauge.
2. "Red" hydraulic system contents gauge.
3. No. 1 engine oil temperature and pressure gauge.
4. No. 1 engine fuel thermometer.
5. No. 2 engine oil temperature and pressure gauge.
6. No. 2 engine fuel thermometer.
7. No. 3 engine oil temperature and pressure gauge.
8. No. 3 engine fuel thermometer.
9. No. 4 engine oil temperature and pressure gauge.
10. No. 4 engine fuel thermometer.
11. "Blue" hydraulic system contents gauge.
12. "Green" hydraulic system contents gauge.
13. No. 1 engine fuel flowmeter.
14. No. 1 engine fuel heater switch.
15. No. 1 engine "fuel filter blocked" warning lamp.
16. No. 1 engine low fuel pressure warning lamp.
17. Port common/independent feed magnetic indicator.
18. No. 2 engine low fuel pressure warning lamp.

Fuel Panel (*Contd.*)

19. No. 2 engine fuel flowmeter.
20. No. 2 engine fuel heater switch.
21. No. 2 engine "fuel filter blocked" warning lamp.
22. Cross-feed cock magnetic position indicator.
23. No. 3 engine fuel flowmeter.
24. No. 3 engine fuel heater switch.
25. No. 3 engine "fuel filter blocked" warning lamp.
26. No. 3 engine low fuel pressure warning lamp.
27. Starboard common / independent fuel indicator.
28. No. 4 engine low fuel pressure warning lamp.
29. No. 4 engine fuel flowmeter.
30. No. 4 engine fuel heater switch.
31. No. 4 engine "fuel filter blocked" warning lamp.
32. Port No. 4 tank isolation indicator.
33. Port No. 4 tank fuel feed indicators.
34. Port No. 3 tank isolation indicator.
35. Port No. 3 tank fuel feed indicators.
36. Port No. 2 tank isolation indicator.
37. Port No. 2 tank fuel feed indicators.
38. No. 1 tank isolation indicator.
39. No. 1 tank fuel feed indicators.
40. Starboard No. 2 tank isolation indicator.
41. Starboard No. 2 tank fuel feed indicators.
42. Starboard No. 3 tank isolation indicator.
43. Starboard No. 3 tank fuel feed indicators.
44. Starboard No. 4 tank isolation indicator.
45. Starboard No. 4 tank fuel feed indicators.
46. No. 1 engine turbine failure warning lamp.
47. No. 2 engine turbine failure warning lamp.
48. No. 3 engine turbine failure warning lamp.
49. No. 4 engine turbine failure warning lamp.
50. No. 1 engine overheat warning lamp.
51. No. 2 engine overheat warning lamp.
52. No. 3 engine overheat warning lamp.
53. No. 4 engine overheat warning lamp.
54. Tank No. 1 jettison warning lamp.
55. Port wing fuel jettison warning lamp.
56. Tank No. 1 jettison warning lamp.
57. Starboard wing fuel jettison warning lamp.
58. No. 4 tank fuel thermometer.
59. Port No. 4 tank fuel gauge.
60. Port No. 3 tank fuel gauge.
61. Port No. 2 tank fuel gauge.
62. No. 1 tank fuel gauge.
63. Starboard No. 2 tank fuel gauge.
64. Starboard No. 3 tank fuel gauge.
65. Starboard No. 4 tank fuel gauge.
66. Rear spar thermometer.
67. Selector—rear spar temperature.
68. Port No. 4 tank fuel contents test button.
69. Port No. 3 tank low-level warning lamp.
70. Port No. 2 tank low-level warning lamp.
71. No. 1 tank low-level warning lamp.
72. Starboard No. 2 tank low-level warning lamp.
73. Starboard No. 3 tank low-level warning lamp.
74. Starboard No. 4 tank fuel-contents test button.

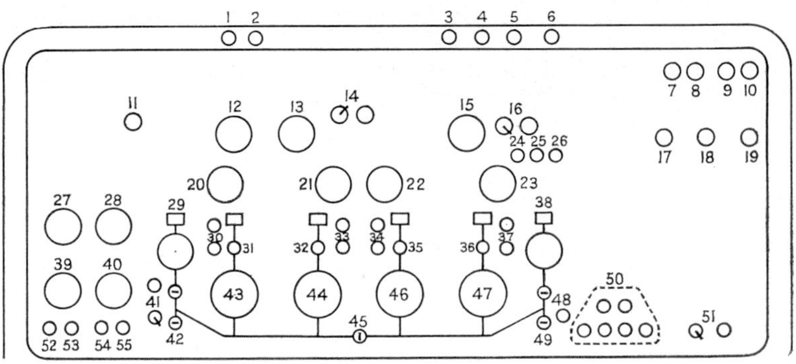

DE HAVILLAND COMET MK. 4

Electrical Panel

1. Forward CO_2 fire extinguisher test circuit fuse.
2. Rear CO_2 fire extinguisher test circuit fuse.
3. Battery 1 volts test fuse.
4. Battery 2 volts test fuse.
5. Stand-by lights fuse.
6. Essential services warning lamp fuse.
7. No. 1 engine fire warning lamp.
8. No. 2 engine fire warning lamp.
9. No. 3 engine fire warning lamp.
10. No. 4 engine fire warning lamp.
11. Busbar failure warning lamp.
12. D.C. voltmeter.
13. D.C. voltage selector switch.
14. D.C. over-volt warning lamp.
15. A.C. voltmeter.
16. Stand-by invertor switch and warning lamp.
17. Port (leading edge bay) fire warning lamp.
18. "Centre section bay" fire warning.

Electrical Panel (*Contd.*)

19. Starboard (leading edge bay) fire warning lamp.
20. No. 1 generator air-outlet temperature thermometer.
21. No. 2 generator air-outlet temperature thermometer.
22. No. 3 generator air-outlet temperature thermometer.
23. No. 4 generator air-outlet temperature thermometer.
24. Stand-by invertor over-volt warning.
25. No. 1 invertor over-volt warning.
26. No. 2 invertor over-volt warning.
27. "Yellow" hydraulic system pressure gauge.
28. "Red" hydraulic system pressure gauge.
29. Battery bank No. 1 ammeter.
30. Port windshield panel de-icing warning lamp.
31. No. 1 generator on/off line magnetic indicator.
32. No. 2 generator on/off line magnetic indicator.
33. Port windshield panel power supply magnetic indicator.
34. Starboard windshield panel power supply magnetic indicator.
35. No. 3 generator on/off line magnetic indicator.
36. No. 4 generator on/off line magnetic indicator.
37. Starboard windshield panel de-icing magnetic indicator.
38. Battery bank No. 2 ammeter.
39. "Blue" hydraulic system pressure gauge.
40. "Green" hydraulic system pressure gauge.
41. Battery bank No. 1 isolation switch and magnetic indicator.
42. Ground supply magnetic indicator.
43. No. 1 generator ammeter.
44. No. 2 generator ammeter.
45. Busbar split magnetic indicator.
46. No. 3 generator ammeter.
47. No. 4 generator ammeter.
48. Battery bank No. 2 isolation switch and magnetic indicator.
49. Ground supply magnetic indicator.
50. Domestic services on/off magnetic indicators.
51. P.E. contactor warning lamp and reset switch.
52. "Blue" hydraulic pump No. 1 engine flow indicator lamp.
53. "Blue" hydraulic pump No. 3 engine flow indicator lamp.
54. "Green" hydraulic pump No. 2 engine flow indicator lamp.
55. "Green" hydraulic pump No. 4 engine flow indicator lamp.

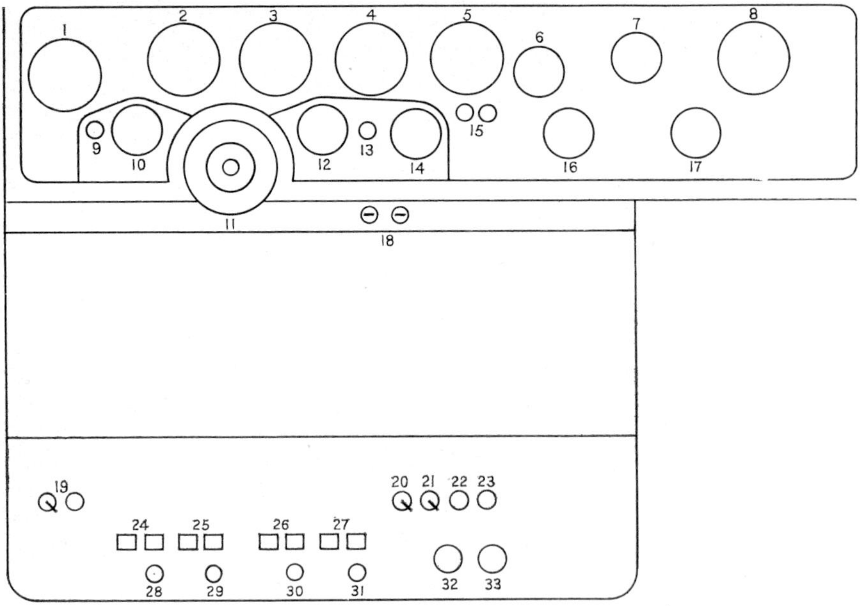

DE HAVILLAND COMET MK. 4

Pressure, Temperature and De-icing Panel

Key to Pressure, Temperature and De-icing Panel

1. Airport altitude selector.
2. Mass flow indicator: No. 1 and No. 2 engines.
3. Mass flow indicator: No. 3 and No. 4 engines.
4. Cabin Rate-of-climb indicator.
5. Differential pressure gauge.
6. Cabin height indicator.
7. Cabin thermometer.
8. Centre section bay thermometer.
9. Port wing de-icing duct over-heat warning lamp.
10. Port wing de-icing duct thermometer.
11. Manual de-pressurization control.
12. Starboard wing de-icing duct thermometer.
13. Starboard wing de-icing duct overheat warning lamp.
14. Tail de-icing duct thermometer.
15. Fuse holders.
16. Cabin air inlet (Port) thermometer.
17. Cabin air inlet (Starboard) thermometer.
18. Internal and external start magnetic indicators.
19. Port wing and engine bay temperature control switches.
20. Starboard engine bay vent intake heaters switch.
21. Starboard wing de-icing duct temperature control switch.
22. Port pitot mast de-icing switch.
23. Starboard pitot mast de-icing switch.
24. No. 1 engine de-icing valves open/shut magnetic indicators.
25. No. 2 engine de-icing valves open/shut magnetic indicators.
26. No. 3 engine de-icing valves open/shut magnetic indicators.
27. No. 4 engine de-icing valves open/shut magnetic indicators.
28. No. 1 engine de-icing valve intermediate position indicator.
29. No. 2 engine de-icing valve intermediate position indicator.
30. No. 3 engine de-icing valve intermediate position indicator.
31. No. 4 engine de-icing valve intermediate position magnetic indicator.
32. Ammeter for item 22.
33. Ammeter for item 23.

GENERAL REQUIREMENTS FOR AIRCRAFT INSTRUMENTS

The British Standards Institution has laid down general requirements for electrical equipment and indicating instruments in British Standard G.100 of April 1952. This specification is summarized in Appendix II, but mention must be made of the fact that G.100 is under revision at the present time and any figures quoted in the Appendix should not be used without reference to the B.S.I. Other specifications relating to particular aircraft instruments are also available from the B.S.I.

TERMINOLOGY

As with other technical fields, instrumentation has its own terminology, and a discussion with instrument engineers would probably show that a number of expressions had, through common usage, lost some of their true meaning. It is worth while at the outset to define some of these terms.

1. *Accuracy*. This implies truth; that is, an instrument is said to be accurate if it agrees with a true measure. This, in turn, implies comparison and it is necessary at regular intervals to verify the accuracy of an instrument by comparing it with a master instrument under specified test conditions and noting how closely the readings of the instrument under test compare with

those of the master instrument. The master instrument may not be the ultimate form of measurement, but merely a convenient method of reproducing a value which compares with the ultimate value. An illustration of this point is given by a micrometer which is used in engineering workshops to measure length. Any wear on the micrometer will introduce inaccuracies in measurement, and so standards (slip gauges) are available for the periodic checking of micrometers. The slip gauges, however, are not the ultimate standard of measurement; one such ultimate standard consists of a metal block on which are inscribed two marks one yard apart. The block is housed under controlled temperature conditions, and for safe-keeping, at the National Physical Laboratory. (Although this is the ultimate standard it is a most inconvenient means of checking micrometers!) Thus, when issuing a verdict on accuracy of the micrometer a possible loss in accuracy in the slip gauges should be borne in mind. In practice, therefore, it is useful to define, and periodically check on, the accuracy of the sub-standard, or master, equipment.

2. *Precision.* This term means, literally, exactness. For example, taking two instruments *A* and *B* and checking them against a sub-standard may show an error in both of, say +1 per cent, which gives them both the same degree of accuracy. A subsequent check against the sub-standard may show that instrument *A* now has $+\frac{1}{2}$ per cent error while instrument *B* again shows +1 per cent. In such a case, instrument *A* would be said to have the higher accuracy, but instrument *B* the greater precision.

A precision instrument can be defined, then, as an instrument in which (test conditions remaining constant) the reading obtained is repeated over and over again. Expressed in another way, it is an instrument in which the repeatability of reading is exact.

3. *Sensitivity.* The sensitivity of an instrument is a measure of the ratio of output to input. As an example, a pilot's altimeter responds to a barometric pressure change of 1,000 ft. by moving a pointer through one complete revolution of the dial, while a navigator's altimeter, for the same pressure change, will move the instrument pointer through only one-tenth of a revolution of the dial. The pilot's instrument therefore has the greater output for a given input and has the higher sensitivity.

4. *Calibration.* The calibration of an instrument is the method of expressing physical or electrical units on the dial. For example, the dial of a pressure gauge may be calibrated in pounds per square inch, kilograms per square centimetre, inches of water, and so on.

5. *Calibrate.* To calibrate an instrument is to adjust the mechanism to obtain the required accuracy of reading of the instrument.

6. *Error.* Instrument error is the difference between the instrument reading and the true reading, and is shown when checking the instrument readings against a sub-standard (refer also to *Accuracy*). The difference between

VICKERS VISCOUNT

Instrument Panels (see diagram overleaf)

1. Inter-engine and cross-feed cocks, switches and indicators.
2. Aileron trim switches and indicator.
3. Door seal pressure gauge.
4. Landing gear position indicator.
5. Generator failure warning lamps.
6. Hydraulic "pump" pressure gauge.
7. Hydraulic "system" pressure gauge.
8. Brake pressure gauges.
9. Three-axis trim indicator.
10. Compass light dimmer.
11. Weather radar indicator.
12. "Compass only" switch.
13. Auto-pilot indicator lamps.
14. Clock.
15. A.D.F. V.O.R. change-over switch.
16. Omni-bearing magnetic indicator.
17. Spare fuses.
18. Auto-pilot caging switch.
19. Airspeed indicator.
20. Marker indicator lamps.
21. Altimeter.
22. R.P.M. indicators.
23. Torque meters.
24. Propeller low-stop-removed warning.
25. Propeller below-low-stop warning.
26. Engine oil low pressure warning.
27. Propeller low stop test lamp.
28. Low pressure fuel cock indicators.
29. Engine fire warning lamps.
30. Magnetic compass.
31. Auto-pilot sequence switch.
32. A.D.F. receiver controller.
33. Rudder horn de-icing switch and warning lamp (when fitted).
34. Pitot head heaters switch and indicators.
35. Airframe de-icing switches and overheat warning lamps.
36. Airframe de-icing temperature gauge and selector switch.

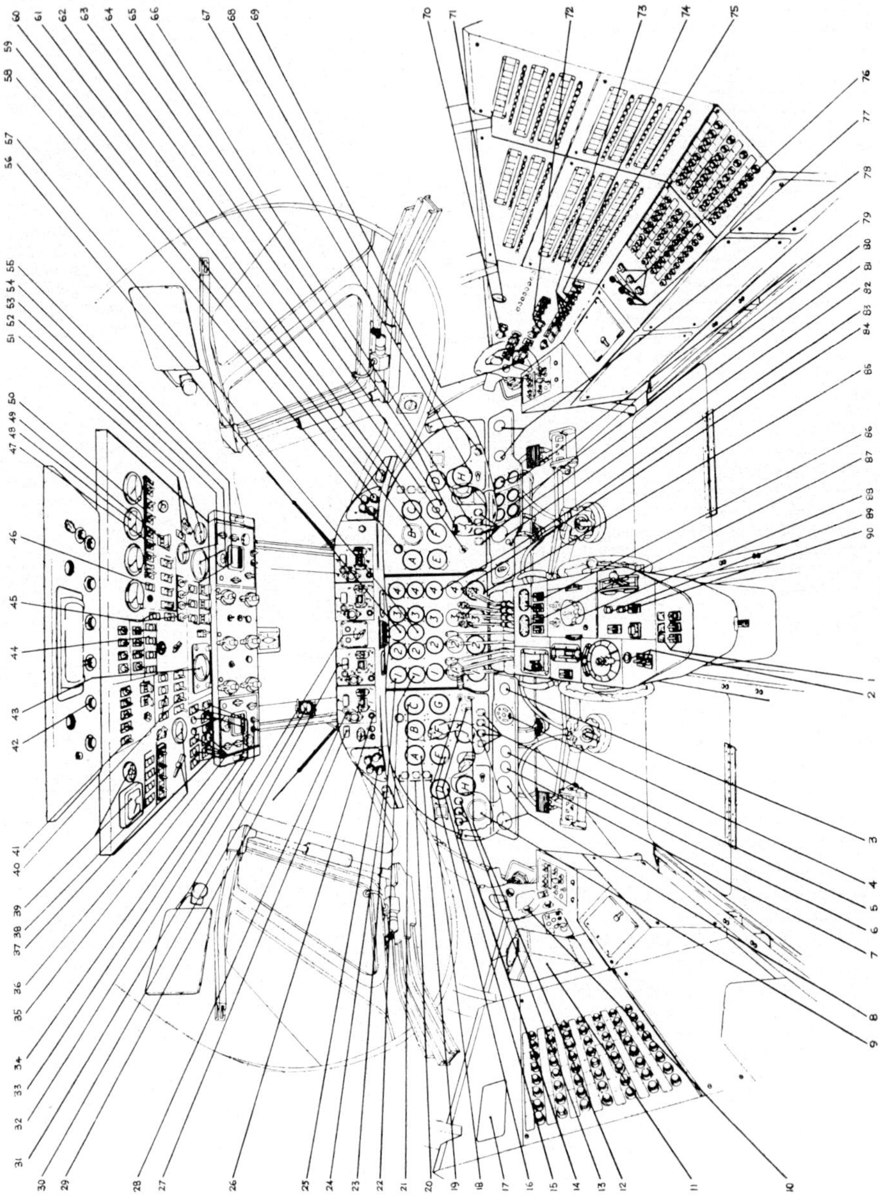

Vickers Viscount Instrument Panels (*Contd.*)

37. Power plant anti-icing switches.
38. Power anti-icing cycling lamps.
39. Power plant anti-icing test ammeter and test selector switch.
40. Fuel filter de-icing switches and icing warning lamps.
41. Panel and cockpit lights change-over switch.
42. Dashboard white lights dimmer.
43. Cabin pressure controller.
44. Fuel low-pressure warning lamps.
45. Starter master switch and start cycle indicator lamp.
46. Ground power warning lamp.
47. Master power switch.
48. Generator volt-ammeters.
49. Invertor change-over switch.
50. Generator master switches.
51. Doors warning selectors switch and warning lamp.
52. Cabin differential pressure gauge.
53. Cabin altimeter.
54. Cabin rate-of-climb indicator.
55. Compressor supply failure warning.
56. Outside air temperature gauge.
57. Flaps position indicator.
58. Invertor failure indicator.
59. Propeller high-stop-removed warning.
60. Gyro horizon unit.
61. Propeller low-stop-removed warning.
62. Master direction indicator.
63. A.D.F. V.O.R. change-over switch.
64. Rate-of-climb indicator.
65. Auto-pilot instinctive cut-out.
66. Radio magnetic indicator.
67. Turn-and-slip indicator.
68. Compass caging switch.
69. Inspection lights warning lamp.
70. Airframe de-icing jet-pipe flap valve lamps.
71. Cabin temperature gauge.
72. Slipper tanks low pressure warning.
73. Fuel contents gauges switches.
74. Water-methanol contents gauges switches.
75. Slipper tanks gauges switches.
76. Windshield cycling indicator lamps.
77. Busbar power indicators.
78. Slipper tanks contents gauges.
79. Water-methanol contents gauges.
80. Fuel contents gauges.
81. Water-methanol indicator lamps.
82. Fuel totalizer and push-button.
83. Turbine gas temperature gauges.
84. Flowmeters.
85. Oil pressure/temperature gauges.
86. Fuel datum trimming switches.
87. Fuel datum trimming indicators.
88. Auto-pilot controller.
89. Flap circuit-breaker-tripped warning lamp.
90. Flap emergency engage indicator lamp.

---

the two is expressed as a positive or negative error, depending on whether it is greater or less than the reading of the sub-standard.

7. *Correction.* The correction of an instrument is the amount required to be added (algebraically) to the reading to obtain the true value. If a master instrument reads 100 lb. per sq. in. (p.s.i.) and the instrument under test reads 102 p.s.i. then the error is plus 2 p.s.i. To apply correction, minus 2 p.s.i. must be added to the reading of the instrument under test; that is, the correction is minus 2 p.s.i. Therefore, error and correction are equal in magnitude but opposite in sign.

8. *Tolerance.* The tolerance of an instrument is the amount by which it may be permitted to depart from the true reading and still perform its function. Tolerance is "permissible error" or "allowable error".

While an instrument should be designed, manufactured and maintained to be as accurate as possible, it is permissible, for economy alone, to allow some departure from the standard. For example, a cheap clock keeps sufficiently good time for household purposes if it is accurate to within two minutes per day, whereas a timekeeper's watch should be accurate to within a fraction of a second per week. There will obviously be a vast difference in cost between the two, but household functions need not be timed by a timekeeper's watch.

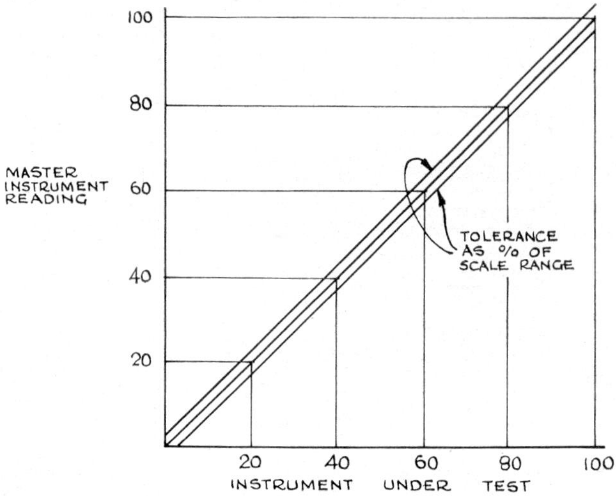

FIG. 1.1.

Inaccuracies in instruments are therefore tolerable, and the amount of tolerance depends on the purpose for which the instrument is required. The tolerance, or permissible error, is usually quoted in one of three ways: (*a*) as a percentage of scale range; or (*b*) as a percentage of observed reading; or (*c*) as a given amount on either side of a scale reading.

(*a*) When expressed as a percentage of scale range the tolerance can be misleading in that a tolerance of 2 per cent of scale range on a 0–100 p.s.i. pressure gauge means ±2 p.s.i. at the 100 p.s.i. point and also ±2 p.s.i. at the 10 p.s.i. point. In effect, at the 100 p.s.i. point the tolerance represents a 2 per cent

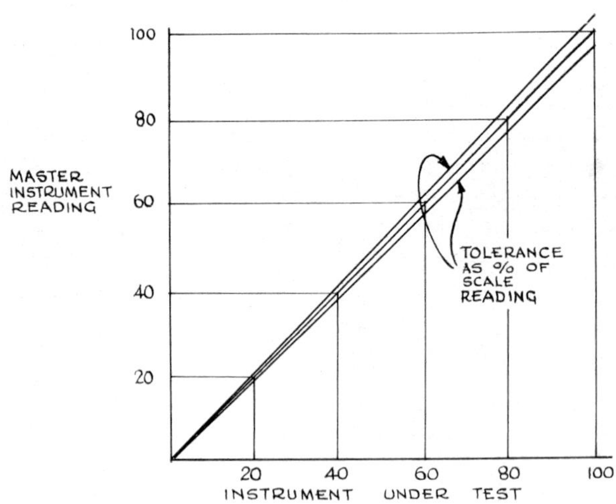

FIG. 1.2.

error in reading, while at the 10 p.s.i. point it represents a 20 per cent error in reading (see Fig. 1.1).

(*b*) When expressed as a percentage of observed reading, the instrument accuracy improves at the lower scale readings. For example, a tolerance of 2 per cent of observed reading would mean an error of ± 2 p.s.i. at the 100 p.s.i. point, but an error of only $\pm 0{\cdot}2$ p.s.i. at the 10 p.s.i. point (see Fig. 1.2).

(*c*) When expressed as a given amount on either side of a scale reading, doubt can only arise when the given amount is not constant throughout but varies from point to point on the range (e.g. ± 1 p.s.i. at the 10 p.s.i. point

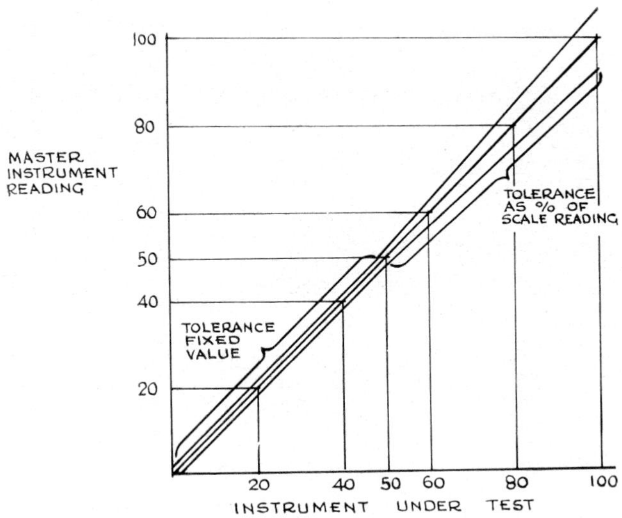

FIG. 1.3.

and ± 5 p.s.i. at the 100 p.s.i. point). In some instruments a combination of (*a*) and (*c*), or of (*b*) and (*c*), may be given. For example, in a 100 p.s.i. gauge the tolerance may be expressed as ± 1 p.s.i. over the range from 0 to 50 p.s.i. and ± 2 per cent of the observed reading over the range from 50 to 100 p.s.i. This is a combination of (*b*) and (*c*) and defines rigidly the practical measureable accuracy over the lower half of the range (see Fig. 1.3).

9. *Hysteresis*. The hysteresis or "lag" of an instrument is the difference between observed readings at specified points when checking from zero to maximum and then back from maximum to zero. This is taken by plotting a curve of errors noted with, say, increasing pressure, and errors taken with decreasing pressure. The difference between the two curves represents the hysteresis or "lag" of the instrument.

CHAPTER 2

General Construction of Instruments

In general there are three separate sub-assemblies which go to make up the final instrument whether of the mechanical or electrical type. These are: (*a*) the sensitive element which detects a mechanical or electrical change; (*b*) the mechanism by which the change is conveyed to a pointer; (*c*) the dial over which the pointer moves to indicate the change.

MECHANICAL INSTRUMENTS

In many mechanical instruments measuring small physical changes the sensitive element is a capsule, diaphragm or bellows unit. These are pressure-sensitive elements used, for instance, in the airspeed indicator, altimeter, rate-of-climb indicator, boost gauge, pressure warning switches and so on. The pressure changes measured by these devices range from a few millimetres of water gauge to many pounds per square inch. Higher pressures up to thousands of pounds per square inch are usually measured by Bourdon-tube instruments which will be described later.

The choice of element depends on the use to which it is to be put and on the pressure to be measured. In general, a capsule or diaphragm will be used to measure pressures of as low as 20 mm. of water, while bellows are used to measure pressures of, say, 200 lb. per sq. in.

The two most common materials for making capsules and diaphragms are phosphor bronze and beryllium copper. The latter material is the stronger

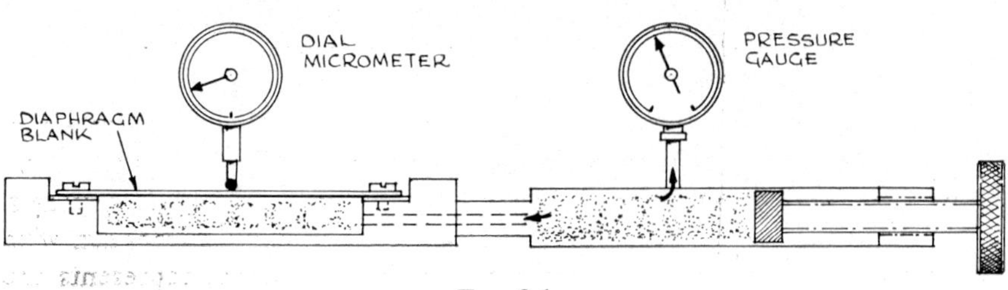

FIG. 2.1.

and is used in the sensitive types of diaphragms and capsules. Bellows are usually made mainly from malleable brass. The thickness of the material varies according to the sensitivity of readings required from the particular instrument. The rate-of-climb indicator, for instance, has a capsule formed of two diaphragms which may be only 0·002 in. thick and measures pressures of a few mm. of water. A high-range airspeed indicator may employ a capsule formed of diaphragms 0·005 in. thick, while a bellows measuring 40 lb. per sq. in. will be in the region of 0·004 in.

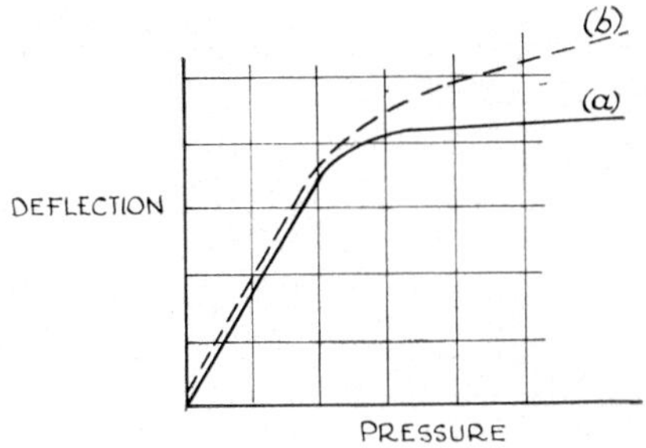

FIG. 2.2.

Characteristics of Capsules and Diaphragms. The characteristics of capsules and diaphragms vary with: (i) thickness of material; (ii) area of capsule or diaphragm; (iii) type of material; (iv) number and depth of corrugations. It is fairly obvious why the thickness, area and type of material should govern the characteristic, and only the fourth point need be considered. Imagine a circular blank cut out of a strip of metal of a certain nominal thickness. This can be mounted in a fixture and a pressure applied while a dial micrometer is used to measure the diaphragm deflection (see Fig. 2.1). If readings of deflection and pressure are obtained a curve might result as in Fig. 2.2(*a*). This indicates a linear characteristic initially, but rapidly falls off until large increases in pressure produce very little deflection.

Now take the same diaphragm and form a corrugation towards the edge of the blank so that a cross-section results, as shown in Fig. 2.3. It will be found on reapplying the pressure test, that the pressure deflection curve has flattened a little, giving a linear characteristic over a greater range of pressure as in Fig. 2.2(*b*). If the diaphragm is now formed with a number of corrugations and the depth of corrugation varied, it will be found in general that as the number and depth of corrugations increases so a more linear pressure deflection characteristic is obtained. It is possible therefore to obtain a limited control of capsule or diaphragm characteristic by varying the thickness of the diaphragm blank and the number and depth of corrugations.

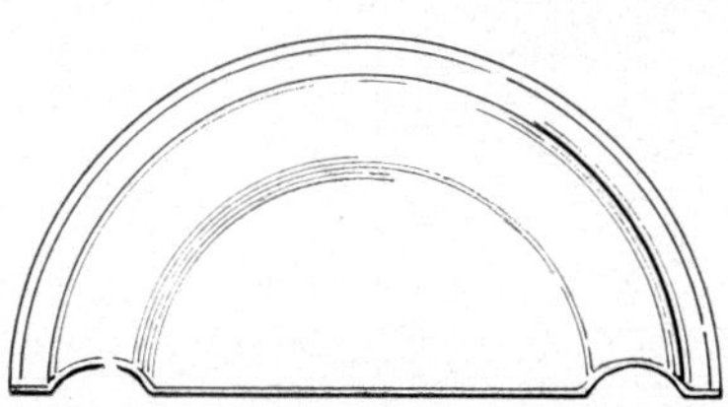

FIG. 2.3.

An altimeter capsule may be formed of two phosphor-bronze diaphragms each 0·007 in. thick with a small number of corrugations. In the type mentioned, the completed capsule is then evacuated to a pressure of about 0·05 mm. of mercury

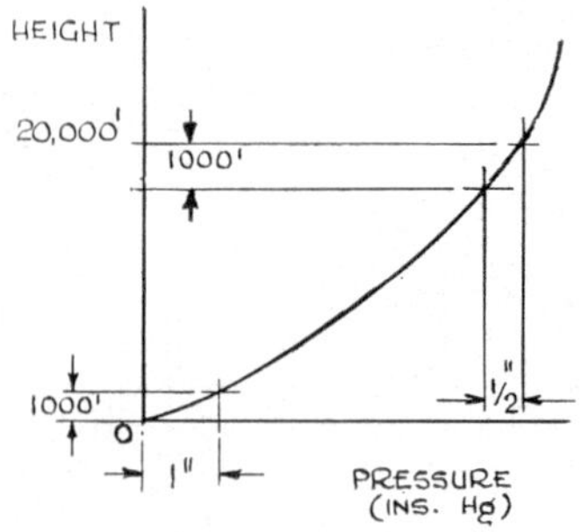

FIG. 2.4.

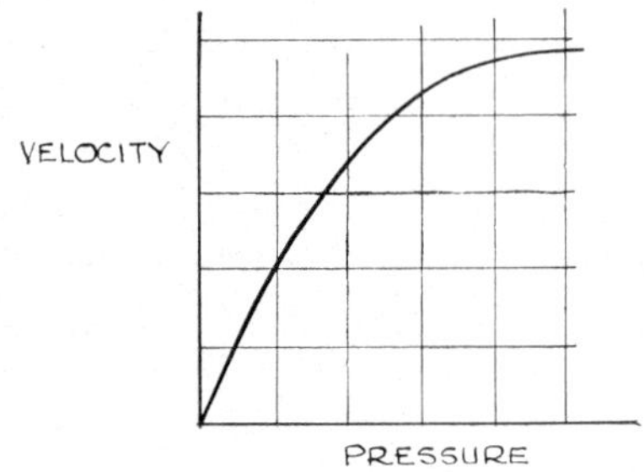

FIG. 2.5.

(Hg) and sealed off. This means that under normal atmospheric pressure the capsule tends to collapse and is only prevented from so doing by the spring tension of the capsule itself. With increasing altitude the pressure decreases and the force acting on the capsule is reduced; therefore the springy nature of the capsule causes it to open out by an amount proportional to the pressure change.

A height/pressure curve would show a non-linear characteristic, as in Fig. 2.4. Because a linear scale on the instrument is required, the capsule characteristic is controlled to give it a small movement at the low end of the scale and a large movement at the high end, enabling a reasonably linear height/deflection characteristic to be obtained. Any remaining non-linearity can be removed by adjustment of the magnifying mechanism. In some instruments there may be two or more capsules to give more power to the movement.

The characteristic required from the airspeed capsule is the reverse of that required for the altimeter. In this instance the capsule measures a pressure which varies as the square of the forward speed of the aircraft. A velocity/pressure curve would be as shown in Fig. 2.5. The airspeed indicator capsule is therefore controlled to give a characteristic which provides greater sensitivity to pressure at the low end of the scale, but less sensitivity at the high end. If this characteristic is used in conjunction with the appropriate type of magnifying mechanism a linear scale will be obtained.

One manufacturer uses a capsule of beryllium copper for low airspeed measurement, each diaphragm being in the region of 0·002 in., while for high airspeed ranges a phosphor-bronze capsule of 0·005 in. is used. Again, the finished instrument may have more than one capsule for greater sensitivity and power.

The rate-of-climb indicator is an extremely sensitive instrument and the capsules used in the instrument are correspondingly sensitive. One manufacturer uses a capsule formed of two diaphragms of beryllium copper each

0·002 in. thick, while another manufacturer uses rolled brass 0·0015 in. thick. The pressures measured by the capsule are quite low (of the order of a few mm. of water only) and are positive and negative depending on whether the aircraft is descending or ascending. The pressure/deflection curve is usually controlled to give a characteristic which is linear over the required pressure range (see Fig. 2.6).

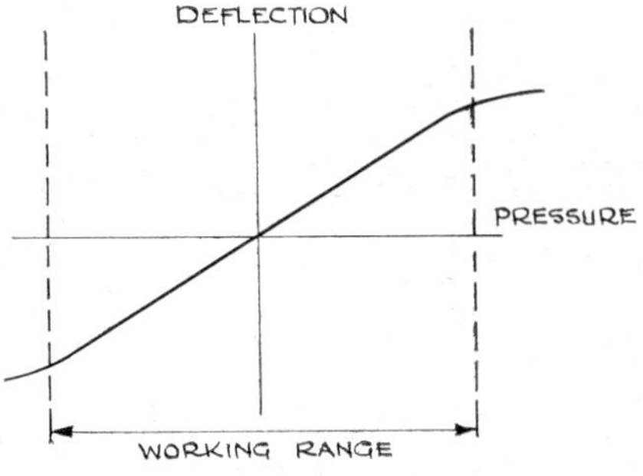

FIG. 2.6.

Characteristic of Bellows. The desired characteristic in a bellows unit is one which gives a linear pressure/deflection curve. Generally, the bellows is not used alone to measure pressure but is restrained by a spring which is the overriding factor in producing the linear characteristic. Nevertheless, even in the more sensitive types of bellows-operated instruments, where the value of the bellows may approach that of the spring, a linear characteristic is found. The thickness of the bellows material is increased for increasing pressure measurement, while the diameter of the bellows is reduced with increasing pressure measurement.

A bellows unit acts as a spring and will vary its length as the pressure changes. However, if an unrestricted bellows is subjected to a high pressure it is possible that the elastic limit of the material would be reached and the bellows would not return to its natural spring length on removal of the pressure. One means of overcoming this is to restrain the bellows with a spring,

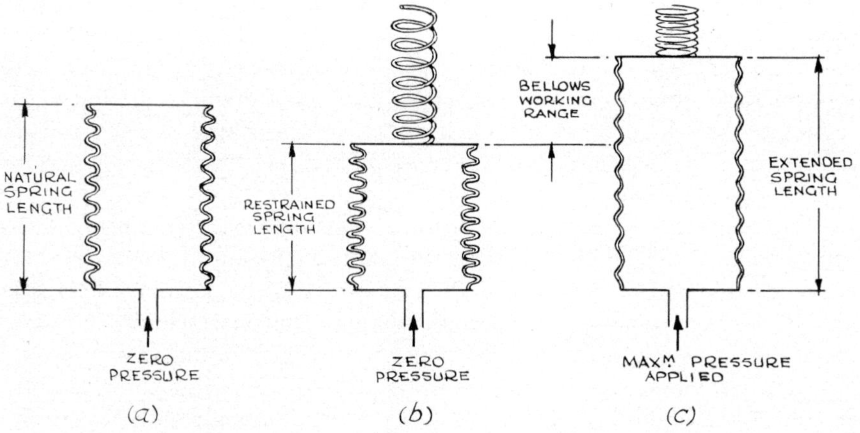

FIG. 2.7.

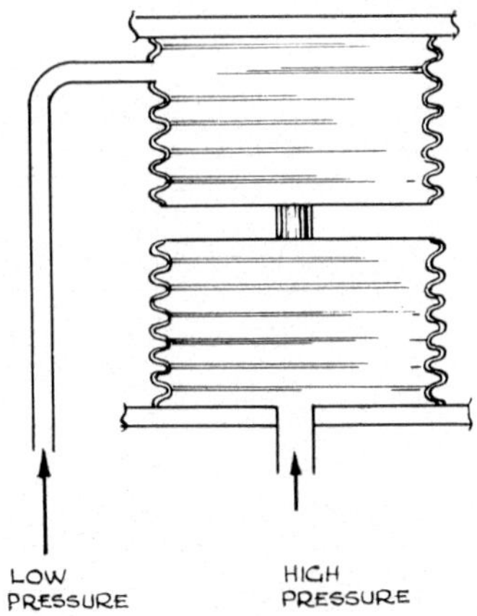

FIG. 2.8.

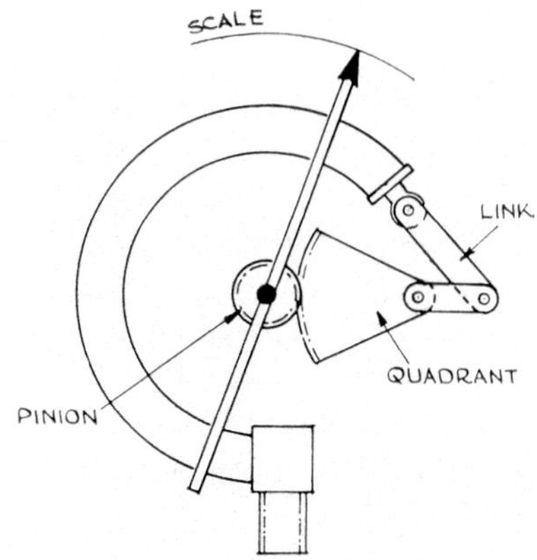

FIG. 2.9.

so that at rest, with no pressure applied, the overall length of the bellows is less than its natural spring length (see Fig. 2.7). As pressure is applied, the bellows returns to its natural spring length before opening out to indicate further pressures. The bellows then operates on either side of its free or natural spring length.

The differential bellows unit comprises two similar bellows units which are arranged in such a way that they oppose each other (Fig. 2.8). If the two bellows are each carrying a different pressure the resulting movement is a measure of the differential pressure. The resulting movement occurs at the centre of the two units; nevertheless, with a suitable mechanism it is possible to transmit this movement to a dial calibrated in terms of differential pressures.

In these notes on diaphragms, capsules and bellows the underlying principles have been outlined. The manufacture of such items is extremely specialized and two manufacturers, if asked to produce a capsule having a given characteristic, would probably produce capsules quite different as to type and thickness of material, number of corrugations and so on. It appears therefore that there is no precise formula for producing a given capsule (although a generalization can be made), but that this depends on the experience of the designer concerned.

Mechanisms. In mechanically operated instruments the variable quantity measured by the sensitive element is relatively small and the response of the element is correspondingly small. The response must be magnified by a suitable mechanism to enable an accurate reading to be observed on the instrument dial. A simple Bourdon-tube instrument (Fig. 2.9) illustrates this.

The free end of a Bourdon tube responds to pressure changes in the tube by moving by an amount proportional to the pressure change. This movement may be quite small; for example, in one case 0·156 in. for a pressure

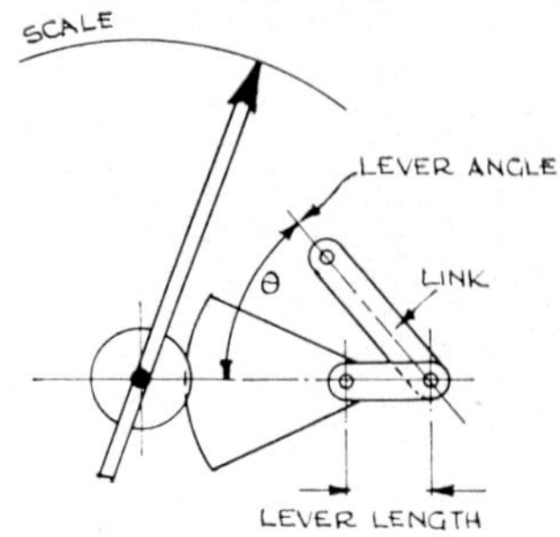

FIG. 2.10.

change of 100 p.s.i., so that a magnifying mechanism consisting of link, quadrant and pinion is employed to move a pointer over a scale covering an arc of about 300 deg. to represent this pressure change.

There are two adjustments on this type of instrument to enable the instrument to be calibrated. These adjustments are: lever-length adjustment; lever-angle adjustment.

Lever-length Adjustment. In the mechanism illustrated in Fig. 2.10 the lever length is the distance between the point of operation of the Bourdon tube and the pivoting point of the lever arm. To simplify the problem, remove the quadrant and pinion so that the pointer is at the end of the lever arm (Fig. 2.11). Movement of the Bourdon tube will result in link movement and operation of the lever and pointer. The pointer movement is proportional to the lever length and if the lever is assumed to be pivoted centrally, then the output movement of the pointer will be equal to the input movement of the link. Now if the lever length is *reduced*, the same input movement will result in a greater ouput movement. Similarly if the length

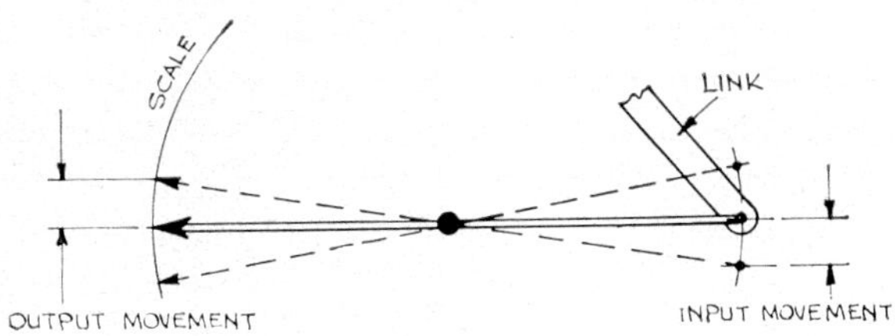

FIG. 2.11.

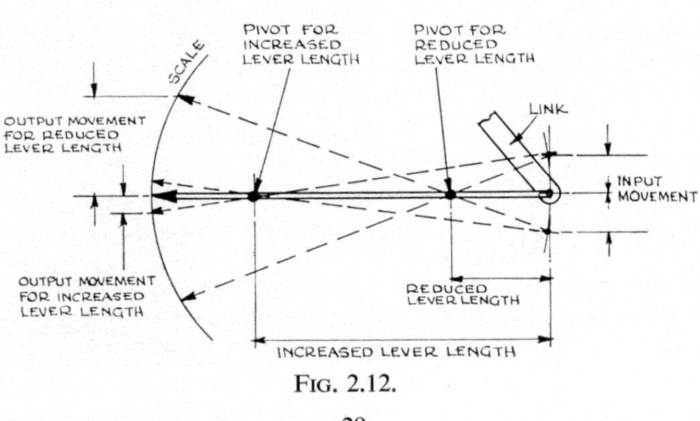

FIG. 2.12.

is *increased*, the output movement is reduced for the same input movement (Fig. 2.12).

Summarizing, a decrease in the lever length means an increase in the reading of the instrument, while an increase in lever length means a decrease in the reading of the instrument for a given input movement.

A typical example of an instrument requiring a lever-length adjustment can be shown by the following table in which is recorded the readings of a pressure gauge when tested against a master instrument:

Master	*Instrument under Test*	*Error*
0	0	0
10	9	−1
20	18	−2
30	27	−3
40	36	−4
50	45	−5
60	54	−6
70	63	−7
80	72	−8
90	81	−9
100	90	−10

that is, the instrument pointer is not moving as far as it should for a given input movement and the instrument needs to be adjusted by *reducing* the lever length. Supposing, however, that the check calibration of an instrument produced the results tabulated as follows:

Master	*Instrument under Test*	*Error*
0	0	0
10	10	0
20	20	0
30	30	0
40	41	+1
50	51	+1
60	62	+2
70	73	+3
80	84	+4
90	95	+5
100	106	+6

A lever-length adjustment to reduce the error at the top end of the scale would result in introducing errors at the bottom end of the scale. In such a case as this a second adjustment to the instrument is usually possible and this is known as the *lever-angle* adjustment.

Lever-angle Adjustment. The lever angle is the angle between the lever which operates the pointer and the link which moves the lever. This lever angle changes as the instrument mechanism responds to a change in the measured variable. The magnification rate changes as the angle changes, and this magnification change can be illustrated by a simple lever-and-link mechanism (Fig. 2.13). Here the link AT moves a lever arm TM through equal increments A_1, A_2, A_3 and A_4, thus moving the lever to T_1M_1, T_2M_2, T_3M_3 and T_4M_4. The angle ATM which started as a right-angle is now an obtuse angle $A_4T_4M_4$ and a glance at the scale will show that the movement from M_3 to M_4 is greater than the movement from M to M_1. Thus the magnification is increasing as an angle increases.

Similarly, if the mechanism is arranged so that the lever angle decreases as the measured variable increases, the reverse is true, that is, the magnification decreases as the lever angle is decreased; e.g. starting with an angle $A_4T_4M_4$, as the angle decreases to ATM the pointer movement decreases for equal increments of the measured variable.

In the instrument whose error table is shown on page 30, the necessary adjustment is one of lever angle, and it is necessary to decrease the starting angle to *reduce* the errors at the top of the scale, while hardly affecting those at the bottom end of the scale. (In making *lever-length* adjustments the *lever angle* may also be changed, but a little practice will suffice to achieve a combination of angle and length adjustments, so providing suitable correction for most instrument errors.)

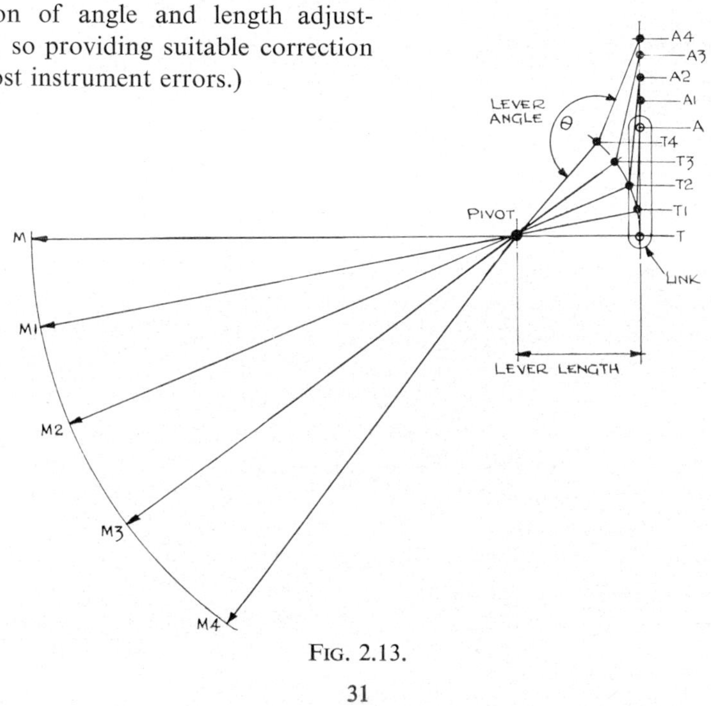

FIG. 2.13.

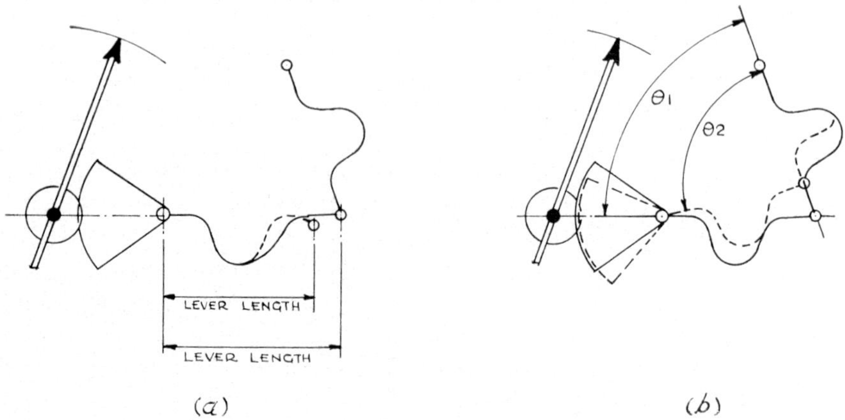

FIG. 2.14.

The initial lever angle will depend on the type of instrument; three typical examples are: (*a*) altimeter; (*b*) airspeed indicator; (*c*) rate-of-climb indicator.

The operating force in the altimeter is change of atmospheric pressure with height. This pressure/height change is not linear, for while the pressure change for 1,000 ft. at sea level is approximately 1 in. Hg, the pressure change for 1,000 ft. at 20,000 ft. may be only $\frac{1}{2}$ in. Hg. Therefore, to enable the altimeter pointer to provide a linear movement with height, the starting angle of the mechanism linkage is acute, so that the pointer "speeds up" as height increases to compensate for the reduced pressure drop for each 1,000 ft. gain in height.

In the airspeed indicator the operating force is an air pressure resulting from the forward speed of the aircraft. The pressure varies as the square of the speed and hence produces a non-linear speed/pressure curve. This is exactly opposite to that of the altimeter, the starting angle for this mechanism being an obtuse angle.

The rate-of-climb indicator measures a differential pressure which is proportional to the vertical speed of the aircraft. This differential pressure will change sign and can be regarded as positive for descent and negative for ascent. The pressure to be measured is very small and it is possible to operate the instrument over the linear portion of the pressure/deflection curve of the sensitive element; thus the starting angle of the mechanism requires to be 90 deg. at zero vertical speed.

What has been said in the foregoing paragraphs would suggest that all non-linearity is removed by suitable choice of lever angle. This is not so and, in practice, much of the linearity of an instrument emanates from the properties inherent in the design of its sensitive element. The lever-angle adjustment allows only the small non-linear errors to be removed.

Types of Linkage Adjustment. It is not possible to show every type of lever-length and lever-angle adjustment but a few typical examples show the principle. Once an understanding of this has been achieved the adjustments on a particular instrument can be found easily. In Fig. 2.14(*a*) both the link and the lever are made with a U-shaped kink in them and the adjustment for lever length or lever angle is primitive, though effective, being simply that the U is closed or opened to achieve the desired result. Should the instrument under-read all the way up the scale, suggesting a lever-length adjustment, the length of the lever is decreased by closing the gap in the U with a pair of pliers. Similarly, if the lever angle is to be increased, the U associated with the link should be closed. These adjustments are shown in Fig. 2.14(*b*).

Fig. 2.15 shows the type of adjustment met with in instruments such as the altimeter and airspeed indicator. The lever-length adjustment is made by loosening the screw in the rocking shaft and pushing the lever arm in or out to decrease or increase the lever length. Notice that the lever angle changes as the lever length is varied. The lever-angle adjustment is made by moving the capsule assembly bodily up or down, thereafter locking it into place when the desired angle is reached.

In the case shown in Fig. 2.16 both link and lever are drilled with a number of holes and the lever length is determined by pinning the link in a given hole along the length of the lever. In the same way the lever angle is determined by the pinning of the lever to the link in a given hole which provides the required angle.

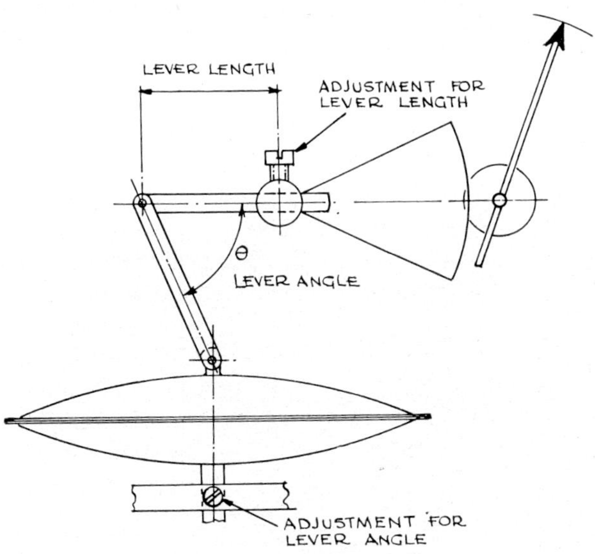

FIG. 2.15.

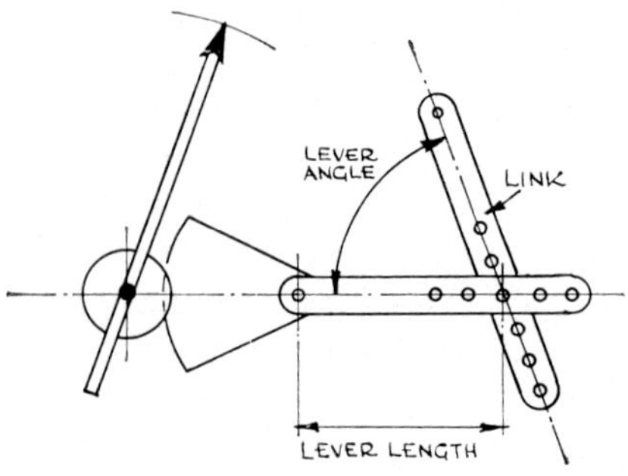

FIG. 2.16.

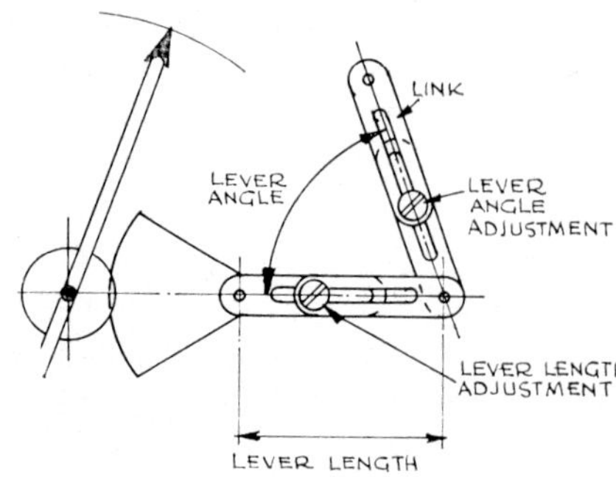

FIG. 2.17.

The example shown in Fig. 2.17 is a more refined version of the adjustment provided in Fig. 2.16, where instead of a number of separate holes, there are two elongated holes with a screw adjustment to determine the length of the lever and similarly the lever angle.

Magnification. The magnification of the operating movement depends on the purpose for which the instrument is required. The altimeter provides a good example of the sort of magnification that can be achieved. In the sensitive type of altimeter the pointer tip travels 7·5 in. in making one revolution of the dial. This pointer is capable of moving 50 times round the dial to indicate a height of 50,000 ft. and therefore the total travel of the pointer is $50 \times 7{\cdot}5 = 375$ in. At 50,000 ft. the sensitive element (a capsule) measuring the altitude has moved only 0·22 in. Thus with an input movement of 0·22 in. and an output movement of 375 in. the magnification ratio is $375 : 0{\cdot}22 = 1700 : 1$ approximately.

The instrument dial enables the change in a variable quantity to be translated into readily understood information as required by the person reading the instrument. In the case of the altimeter, for example, the change measured by the sensitive element is really the change of atmospheric pressure. However, since the relationship between pressure and height is known, it is possible to mark the dial in terms of altitude in feet or metres.

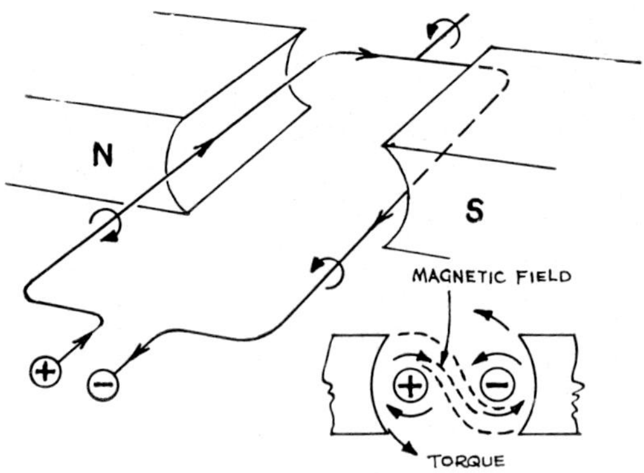

FIG. 2.18.

In general, the calibration marks are made on the dial before fitting it to the instrument. This means that the instrument mechanism will need to be adjusted so that the pointer indicates against successive marks on the dial for appropriate changes in the measured quantity. There are occasions when it is not possible to do this; in such cases, the dial is fitted to the instrument, then changes in the measured quantity are made and the pointer position is noted and marked on the dial. The marks are then engraved on the dial before it goes through the process of painting and so on. It is obvious that in this case the dial is suitable only when assembled to the mechanism with which it was originally calibrated.

ELECTRICAL INSTRUMENTS

In many cases the sensitive element of an electrical instrument is a "moving coil", as shown in Fig. 2.18. A simple moving coil may consist of a rectangular copper-wire loop which is pivoted at each end so that it will turn in a magnetic field. When a current is passed through the rectangular loop a magnetic field is formed round the conductor; this field reacts with the field of the permanent magnet, producing a torque (see Fig. 2.18, inset) which causes the loop to turn in an anti-clockwise direction.

In a practical moving-coil instrument the coil is formed of many turns of very fine wire supported on an aluminium former. The magnetic field produced by a current flowing through a multiple-turn coil will be much stronger than the field from a single turn carrying the same current, and so a greater torque is produced. In a practical instrument the torque of the moving coil is balanced by that of a hairspring, so that the coil deflection is proportional to the current passing through it. Some of the features of a moving-coil instrument are shown in Fig. 2.19.

Ammeters and Voltmeters. Ammeters and voltmeters for use with direct current usually employ the moving-coil principle, and this often gives rise to confusion between the two instruments, because they are basically the same in construction.

An *ammeter* is an instrument which is intended to indicate the flow of current in a circuit without in any way impeding it. The instrument

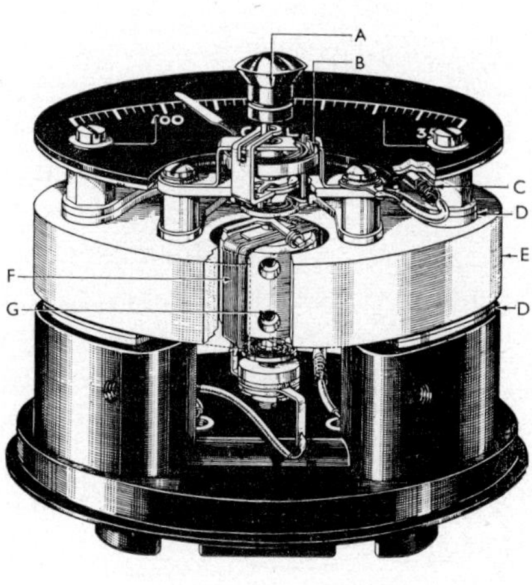

FIG. 2.19. *Moving-coil millivoltmeter showing:* A, *point-setting adjuster;* B, *bi-metal spring forming part of the pointer-control system;* C, *calibrator resistor;* D, *temperature-compensating strips;* E, *permanent magnet;* F, *moving-coil element;* G, *soft-iron core.*

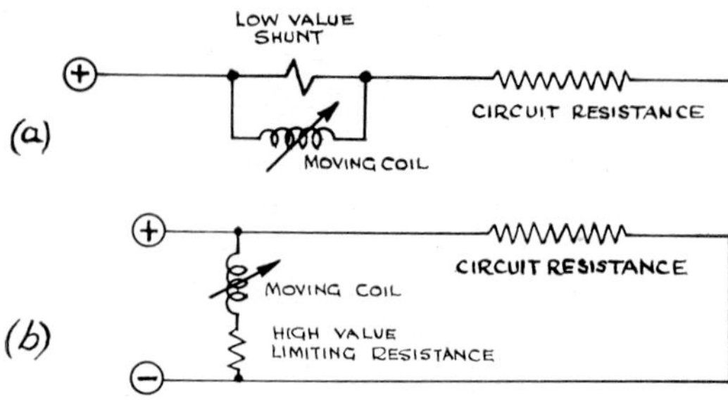

FIG. 2.20.

must be connected "in series" in the circuit whose current is to be measured. The circuit current is governed by the resistance of the appliance connected in the circuit, as well as the voltage applied to the circuit. The moving-coil element itself is very delicate and the pointer will give a full-scale deflection for only a few milliamperes of current, so that when measuring heavier currents the instrument must be shunted. This is done by means of a low-resistance shunt connected across the terminals of the coil as shown in Fig. 2.20*a*.

A *voltmeter* for use on a d.c. circuit may also employ a moving-coil element. Such an element has a low resistance, and the moving coil itself will produce a full-scale deflection of the pointer for only a few milliamperes of current, as previously stated. A voltmeter has to be connected to certain points in a circuit in which there is a comparatively large difference of potential, so that it is necessary to connect a high-value limiting resistor in series with the moving coil, as shown in Fig. 2.20*b*. It will be understood that such a voltmeter is really a current-measuring instrument with the scale calibrated in volts.

Mechanisms. In the case of the simple sensitive electrical element described, there is no mechanism as a rule, the pointer being directly mounted on the coil. However, where added sensitivity of reading is required, it has been known for the coil to be used to drive a fibre gear which meshes with a pinion carrying the pointer. No adjustments, as a rule, can be made to any mechanical gear between pointer and sensitive element. Adjustments that can be made are usually in respect of the strength of the permanent magnet, and this can be done by means of magnetic shunts.

Dials. As in the case of mechanical instruments it is possible to produce a standard dial and adjust the instrument so that the pointer produces correct readings on the dial. However, in instruments of the electrical type it is often the case (more so than with the mechanical instrument) that the dial is calibrated to the movement. It is essential to remember that apparently similar dials may not give the same results when used on different electrical instruments even of the same code and manufacture. Electrical instruments other than moving-coil instruments are used for aircraft instrumentation, but these will be described in detail later, in the appropriate chapters.

CHAPTER 3

Measurement of Pressure

It is possible to measure pressure by means of very simple instruments and, because of this, the problem of pressure measurement is often over-simplified. This chapter deals with pressure measurement and the various instruments used in this form of measurement.

Pressure can be defined as *force per unit area*; e.g. if a force of 10 lb. is applied to an area of 5 sq. in., the pressure = 10/5 = 2 lb. per sq. in. Similarly, if a force of 1 kilogram is applied to an area of 5 sq. cm. the pressure would be 1000/5 = 200 grams per sq. cm. (or 0·2 kg cm.^{2}).

EXPERIMENT IN PRESSURE MEASUREMENT

A simple laboratory experiment is to take a glass U-tube (a tube shaped like the letter U) and partially fill it with a liquid, or with mercury; this will find its own level in each leg of the tube as shown in Fig. 3.1(*a*). Now apply a pressure of 5 p.s.i. to one end of the U tube. With a tube-bore having an area of 2 sq. in. and remembering that pressure = force/area, then force = pressure × area. Therefore a force of 5 × 2 = 10 lb. is acting on the surface of the mercury at (*b*) forcing down the level at one side, so raising the level at the other. Eventually a state of equilibrium will exist and it will be possible to measure the difference in levels (Fig. 3.1*b*). If the difference in levels is *h* inches, the volume of mercury contained in the length of tube is *h* × area = 2*h* cu. in. The density of mercury is given as 0·49 lb. cu. in. so that

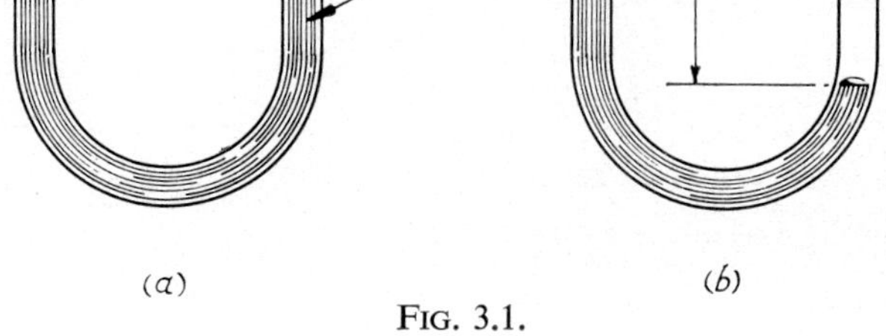

FIG. 3.1.

the weight of mercury in this length h is $2h \times 0{\cdot}49 = 0{\cdot}98h$ lb. Now this weight is creating a force which is balancing the force created by the pressure in the U tube at (*b*), therefore the force represented by $0{\cdot}98h$ lb. = the force represented by 10 lb.

$$\text{i.e. } 0{\cdot}98h = 10$$

$$\text{and } h = \frac{10}{0{\cdot}98} = 10{\cdot}2 \text{ in.}$$

Therefore a height of 10·2 in. of mercury represents a pressure of 5 p.s.i. This experiment can be repeated using glass tubes of differing cross-sectional areas, i.e. by trying the same calculation using tubes of $\frac{1}{2}$ sq. in. area, 1 sq. in., 3 sq. in. and so on. It will be found that the result is independent of the area.

If a pressure of 5 p.s.i. is represented by a difference of levels h of 10·2 in., then 1 p.s.i. would be represented by 10·2/5 = 2·04 in. It is possible therefore to make a scale in which the divisions are 2·04 in. apart and these can be marked in p.s.i. Remember that h is the difference in levels, and its value is obtained by subtracting the lower level from the higher level. Such a device as that described is known as a *manometer* (see Fig. 3.2).

MANOMETER

In Fig. 3.2(*a*) the manometer is shown with levels at zero. In Fig. 3.2(*b*) a pressure has been applied which reduces the level on one side, so increasing that on the other. The difference in levels $h = 6{\cdot}12 - (-6{\cdot}12) = 12{\cdot}24$ in. and the pressure is 12·24/2·04 = 6 p.s.i. However, to simplify the reading of the scale it is usual to arrange for this to be on one side of the manometer only and readings are taken of the levels *at that side*. Taking the previous example (*b*) the level on one side fell by 6·12 in. when the pressure was calculated to be 6 p.s.i. and if the readings are to be taken from this side, then a distance of 6·12 in. on the scale represents 6 p.s.i., and 1 p.s.i. = 1·02 in. (Fig. 3.3). Suppose the manometer is filled with water. Again, a pressure applied to one side will cause a difference in levels until the column of liquid

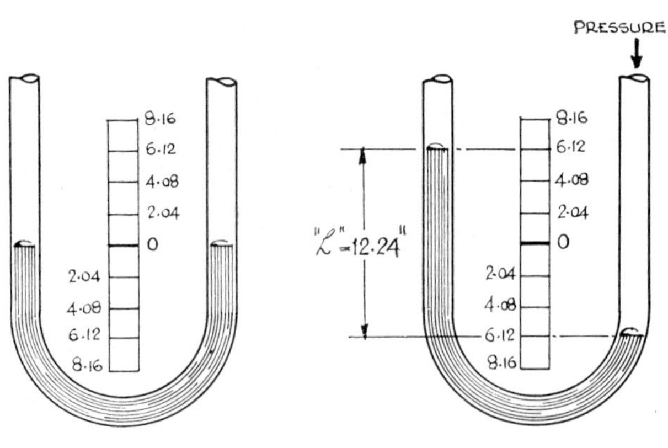

FIG. 3.2.

These then are the forces which tend to straighten the tube, the total deflection of the tube being proportional to the pressure applied. Since section *E* in Fig. 3.8 is anchored to the screwed metal connection, the resultant of the forces on each section causes the maximum movement to occur at section *A* which is, therefore, used as the measuring point. In the case of the Bourdon tube, which showed a change of cross-section of 0·003 in. for a pressure of 100 lb. per sq. in., the total movement at the end of the tube for this pressure was 0·156 in.

DIRECT-READING PRESSURE GAUGES

The direct-reading pressure gauge contains a Bourdon tube, as just described, and is known as a "direct-reading" gauge simply because the pressure being measured is led directly to the instrument.

Among direct reading pressure gauges used in aircraft are numbered oil pressure gauges, hydraulic gauges, fuel gauges, brake gauges and so on. The brake pressure gauge is usually a multi-pressure-reading instrument measuring the air pressure in the braking system of each landing wheel and the pressure available from the source of supply, so that it is customary to see three or four pointers on the same instrument operating over independent scales or occasionally over a common scale. The brake pressure gauge is, however, simply an instrument which houses three or four Bourdon tubes each complete with a separate and independent mechanism all housed in one case. For calibrating, maintenance and testing purposes, each movement is treated separately.

Although both oil and fuel pressure gauges of the direct type are used on aircraft, there is need for a pressure gauge of a different type. It has been shown that the direct type of gauge has the actual medium to be measured brought right up to the connection at the rear of the case and, therefore, filling the Bourdon tube. Now, if a gauge is to measure, say, oil pressure, then a pipe-line must be provided, filled with oil under pressure and connecting the gauge to the engine. If the distance between these is very great, the oil in the pipe-line will thicken, because of its temperature, and cause the instrument to be sluggish. Again, any leak in the pipe-line assembly causes a loss of lubricating oil. While a small leak will not materially affect the functioning of the engine, a fracture of the pipe-line may cause an engine failure. For these reasons the direct type of pressure gauge for fuel and oil has gradually been replaced by an instrument of the transmitting type, either mechanical or electrical.

MECHANICAL-TYPE TRANSMITTING PRESSURE GAUGES

This type of gauge consists of a transmitting unit, a Bourdon-tube indicator and a capillary tube which connects the transmitter to the indicator.

The whole forms a complete unit, and failure of the indicator, capillary tube or transmitter is regarded as a failure of the whole unit. It is not convenient or practical to separate these items, except under factory conditions.

The transmitting unit consists of a pressure-tight box (banjo union) which is bolted, by means of a hollow bolt, to a supply point for the pressure to be measured. Pressure entering the banjo union operates on a capsule which is contained in the union. The capillary tube is a fine-bore ($\frac{1}{2}$-mm. dia.) tube of annealed copper or similar material, which is protected by a cover of waxed cotton braiding. The indicator is of the Bourdon-tube type, in which pressure is conveyed to the tube by a transmitting fluid, and the fluid completely fills the capsule, capillary tube and Bourdon tube. The principle of the transmitting pressure gauge is illustrated in Fig. 3.10.

When pressure enters the banjo union, it operates on the capsule in such a way as to tend to collapse the capsule. This exertion of force by the walls of the capsule creates a pressure on the fluid contained by the capsule. The pressure on the transmitter fluid is felt throughout the length of the capillary tube and at the Bourdon tube, which opens out to indicate the pressure inside it. The movement of the Bourdon tube is therefore a measure of the pressure in the banjo union.

Almost any liquid could be used as a transmitting medium, but the choice of liquid employed is governed by practical considerations; e.g. since the bore of the capillary is very fine, the liquid must be of low viscosity, otherwise the indicator of the instrument would lag behind the pressure at the banjo union. The conditions under which such a gauge may have to work will also influence the type of liquid. For instance, in aircraft work the temperature of the atmosphere in which the instrument is operating (according to I.C.A.N. law) may be as low as $-56{\cdot}5$ deg. C. at 36,090 ft. and therefore a liquid with a low freezing point is required. A liquid often used for this work is ethyl alcohol, which has a low viscosity, and a freezing point around -115 deg. C.

Fracture of the capillary tube means a loss of the liquid in it and a resulting loss of

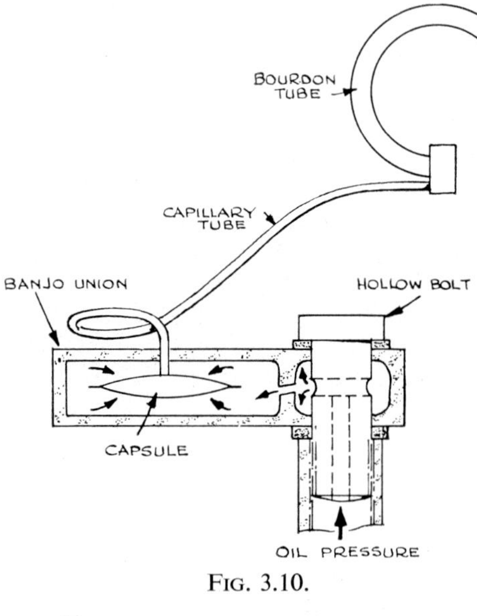

FIG. 3.10.

indication but, unlike the case outlined on page 43, there is no loss of the medium whose pressure is being measured. Thus, in the case of an engine-oil pressure gauge, there is no danger to the engine.

Maintenance. Little maintenance is required on direct-reading and transmitting types of instruments, but if an inspection is being carried out, the main items can be summarized as follows.

(*a*) Check for wear of pivot holes and, if excessive play is evident, close up the holes by means of a hollow punch.

(*b*) Check tension of hairspring. Incorrect setting of the hairspring together with, or independent of, too much play in pivot holes, causes oscillation of the pointer about a reading if the instrument is subject to vibration.

(*c*) Check that the choke, if fitted, is free from foreign matter.

It should be noted that a "choke" is fitted in the connection at the back of some direct-reading instruments to ensure reasonably steady readings, even though the pressure may be pulsating. In the case of transmitting instruments the capillary tube forms a convenient damping device.

The "choke" or restrictor may be in one of several forms; for example, it may simply be a fine-bore hole in the screwed metal connection of the instrument through which the medium under pressure is conveyed to the Bourdon tube, or it may take the form of a grub screw which is inserted in the screwed metal connection and which is drilled down its centre with a fine drill.

(*d*) Check that no leaks occur from the Bourdon tube when under pressure. If the gauge is of the high-pressure type it is usual to drill a hole in the rear of the instrument case, so that if a leak develops there is no possibility of pressure building up behind the glass front. To exclude dirt from the case the hole is normally covered by a substance which breaks down under a very low pressure.

(*e*) Check that the gear teeth on quadrant and pinion are not excessively worn by the constant vibration to which the instrument is normally subjected.

Calibration. Calibration of the pressure gauge means comparison of the readings of the gauge with a master pressure gauge or standard, and adjustment of the instrument until it agrees with the master. While the U tube, or manometer, is useful for calibrating low-pressure gauges, the test equipment to be used for checking high-pressure gauges is known as a dead-weight tester. This is illustrated in Fig. 3.11, where it will be seen that the tester consists of a cylinder containing oil, a piston and a platform for carrying weights, provision for filling the cylinder, a means of attaching the instrument under test, and a screw plunger which is used to build up a pressure in the system.

The operation is as follows. The plunger is screwed into the cylinder, so creating a pressure on the fluid which tends to push the piston out of the cylinder. By placing weights on the piston a pressure can be built up in the system until it is capable of freely supporting the weights on the platform. This pressure registers on the instrument under test and its accuracy can be verified by comparing the reading with the known weight which rests on the piston; e.g. supposing that the piston has an area of 1 sq. in. and a weight of 5 lb. is placed on top of the piston, this is equal to a pressure of 5 lb. per sq. in. When this weight is freely supported by the oil under pressure (as in Fig. 3.11) the instrument under test should show a reading of this value, i.e. 5 lb. per sq. in.

It will be appreciated that to create a pressure of 1,000 lb. per sq. in. by this means it would be necessary to place a weight of 1,000 lb. on top of the piston. In practice, therefore, it is customary to reduce the area of the piston and reduce the value of the weights accordingly; e.g. a weight of 2 oz. on a piston of $\frac{1}{8}$ sq. in. area is equivalent to a pressure of 1 lb. per sq. in. A pressure of 400 lb. per sq. in. can be achieved by placing a weight of 50 lb. on this piston.

One thing to be considered under these conditions is that the piston and platform have weight, and this must be included when reckoning the total weight supported by the oil pressure. The method of procedure when making adjustments to the instrument under test is outlined on page 29.

VACUUM GAUGE

A vacuum gauge is a negative-pressure gauge measuring pressures below that of the atmosphere. The Bourdon-tube instrument is suitable for measuring negative pressures and its appearance is identical with that of the direct type of pressure gauge.

If the Bourdon-tube gauge is connected to a source of pressure less than that of the atmosphere the tube tends to curl up, the action being the same as

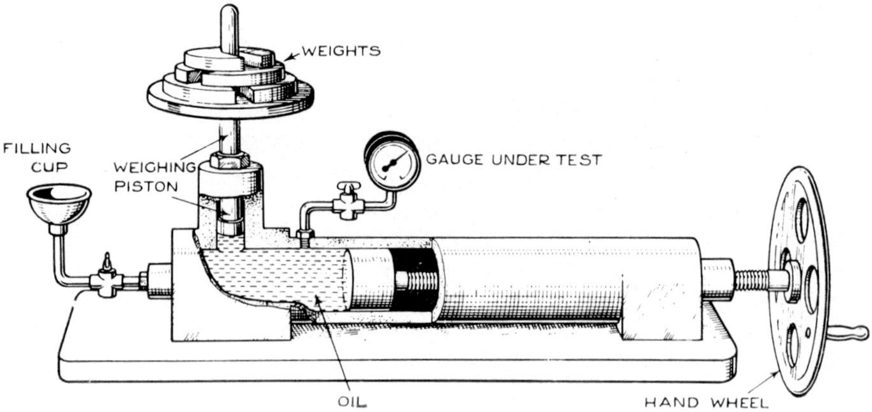

FIG. 3.11.

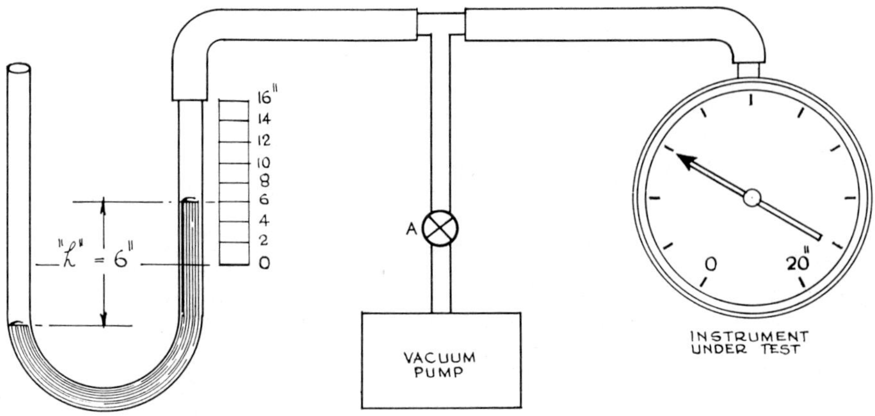

FIG. 3.12.

that of the positive-pressure Bourdon tube on reducing pressure. The indication given by the vacuum gauge is usually in terms of *suction* the units used being inches of mercury (in. Hg). An indication of two inches of suction is therefore equivalent to a negative pressure of 1 lb. per sq. in., i.e. 1 p.s.i. below atmospheric pressure.

For calibration purposes the instrument is connected to a vacuum pump with a mercury manometer connected in the same line (Fig. 3.12). By opening valve *A* the instrument reading is increased and the mercury level changed. The indicator readings should agree with the reading of the mercury level against a calibrated scale.

It must be emphasized that it is the *difference* in the mercury levels that is significant, and care must be taken when reading the manometer scale, as this is fitted against only one leg of the U tube; for example, an increase in one level of $\frac{1}{2}$ in. and a decrease in the other level of $\frac{1}{2}$ in. is read as a difference of levels of 1 in. When reading off a scale connected to one leg of the manometer only, a difference of pressure of 1 in. Hg is represented by a mark $\frac{1}{2}$ in. away from the balanced or zero position.

Adjustments to the mechanism are carried out as for the pressure gauge. Maintenance is the same as for the direct type of pressure gauge.

PRESSURE SWITCHES

There are occasions in pressure measurement when it is not necessary to have a continuous indication of pressure, but rather a signal that pressure has reached a given value. On such occasions a pressure switch can be used in place of the pressure gauge. The pressure switch may contain a single diaphragm or a capsule, and it may signal a rise of pressure to a given value or a fall in pressure to some minimum value. However, in each case the

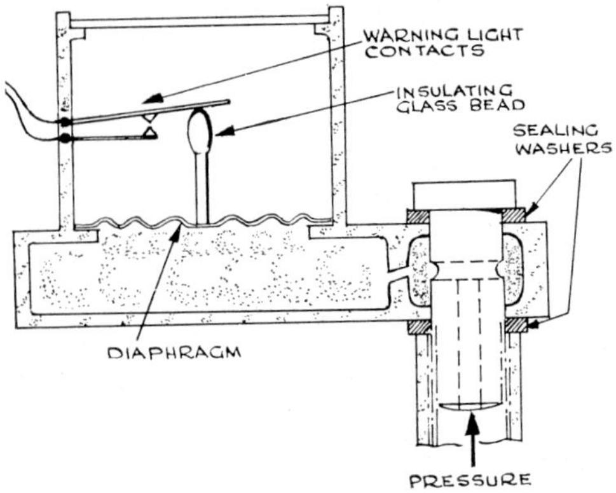

FIG. 3.13.

principle is the same and is shown in Fig. 3.13. The pressure switch is similar to the banjo union of the transmitting-type pressure gauge in that a hollow bolt carries a source of pressure directly to a diaphragm which moves to show the pressure. In this case, however, movement of the diaphragm is conveyed via a pin to a movable electrical contact that provides for the operation of a warning light. The switch may be constructed so as to light the lamp either on reaching a given pressure or when a decreasing pressure is reduced to some minimum value (as in Fig. 3.13).

The range of the switch is governed by the nature and thickness of the diaphragm, while the operating pressure within a given range is controlled by means of a screw adjustment; e.g. a pressure switch has a range of, say, 0–100 lb. per sq. in. and it is required to operate a warning light when a pressure of 75 lb. per sq. in. is reached (Fig. 3.14). The pressure acting on the diaphragm causes movement of the pin so that the lower contact is pushed upwards to meet the upper contact, thus operating the warning light. The contact-adjusting screw is the means whereby the upper contact is adjusted to a point where it meets the lower contact when a pressure of 75 lb. per sq. in. is applied to the switch unit. The operating pin is insulated from the switch by means of a small paxolin pad on the under side of the first contact, or, as in Fig. 3.13, by a glass bead cemented to the pin.

The bush in which the operating pin moves is used also as an overload stop by adjusting it to a suitable height above the diaphragm. The electrical contacts are mounted on a terminal moulding which also carries the contact-adjusting screw. The method of mounting the contacts varies according

FIG. 3.14.

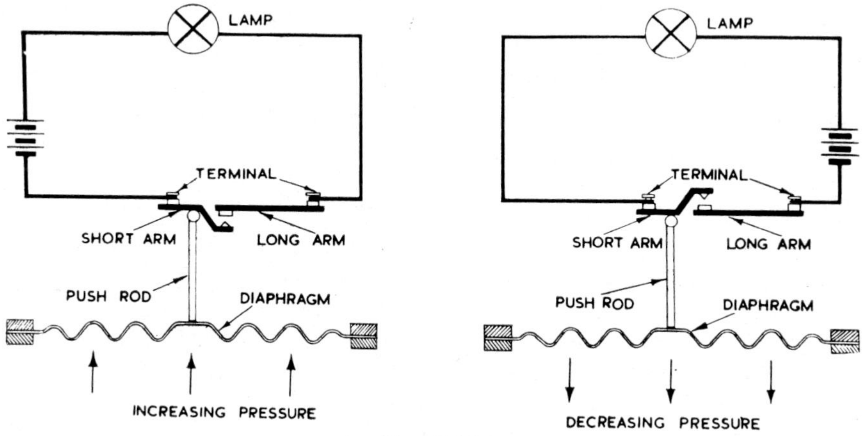

FIG. 3.15.

to the warning to be given; i.e. pressure rising or pressure falling (Fig. 3.15).

Low-pressure and High-pressure Warning Switches. The low-pressure switch is used for measuring pressures up to 100 lb. per sq. in. and is illustrated in Fig. 3.13. The high-pressure switch operates on the same principle, but since it is intended for pressures up to 1,000 p.s.i., it is much more robust. The diaphragm is contained in a pressure chamber which limits its movement. The lower half of the pressure chamber has a screwed union body in place of the hollow banjo connection of the low-pressure switch, but the contact moulding is the same in each case.

The differential pressure switch is slightly different from either the low- or high-pressure switches in that the pressure is measured against some other pressure which is not necessarily atmospheric. The switch has therefore two connections: (*a*) the "high" pressure connection which is of hollow banjo type; and (*b*) the "low" pressure connection which is a screwed union in the side of the switch body. The differential switch has therefore two pressure chambers, one beneath the diaphragm and one formed by the top of the diaphragm and the under side of the terminal moulding. For this reason the moulding is of a more robust nature than that of the low- or high-pressure switch and it must be pressure-tight. The moulding is seated on a rubber ring and screwed down to form an air-tight joint.

The altitude switch is a type of pressure switch containing an evacuated capsule which opens out at increased altitude. The capsule movement is conveyed, via the pin, to the electrical contacts to indicate a given altitude. The switch may operate on reaching a given height with the aircraft ascending, or on reducing height to some minimum level.

The lamps operated by the various switches mentioned are usually of a type with a maximum current of 0·4 ampere. As the lamps are used on 12-volt and 24-volt systems it is necessary to use limiting resistors. It is important to see that, in the case of a fuel-pressure switch, the resistors are of the non-inductive type, so that arcing does not occur at the contact points.

Calibration in the case of the pressure switches consists of nothing more than the setting of the contacts at one particular pressure. In the case of the high, low and differential switches, the setting is effected by connecting the switch to a dead-weight test rig and, after applying the appropriate pressure, adjusting the contact screw until the warning lamp lights or goes out, according to the function of the switch. In some cases the contact-adjusting screw is locked in position by means of a locking unit, but switches of later design have an indexing device which provides its own locking. Setting of the contact points on the altitude switch is effected by placing the switch in a vacuum chamber and reducing the pressure until it is equivalent to that for a given height. The contact points are adjusted accordingly.

The maintenance of these switches is simple. The main points to be noted are as follows. (*a*) Check contact points for "pitting" and, if necessary, replace them with a new set complete with moulding. (*b*) Check that there is insulation between the operating pin and the contacts (either a glass bead or some insulating fabric). (*c*) In the case of the high, low and differential switches, check that there are no leaks in the assembly. This is done by connecting the switch to a dead-weight test rig and applying an overload pressure appropriate to the type of switch being tested.

BOOST GAUGE

The boost gauge is a pressure gauge which measures the pressure of the fuel-air mixture in the inlet manifold of an internal-combustion engine. In this instance the pressure indicated is not that of pressure relative to the atmosphere but the absolute pressure. The reason why the boost gauge measures absolute pressure can best be understood by considering the purpose of this gauge. It will be appreciated that the power of an internal-combustion engine depends, all other things being equal, on the amount of fuel-air mixture forced into the cylinders of the engine. This in turn depends on the pressure difference across the carburettor, which is due to the downward stroke of the piston working in the cylinder on the one hand and atmospheric pressure on the other. In the case of an aircraft engine, which operates at varying altitudes, the power will depend on altitude, since atmospheric pressure decreases with altitude. Eventually a height is reached where the atmospheric pressure is such that the power of the engine will no longer pull the aircraft to a higher altitude. Under these conditions the engine is said to have a ceiling of, say, 15,000 ft., and such an engine is described as a *normally aspirated* engine.

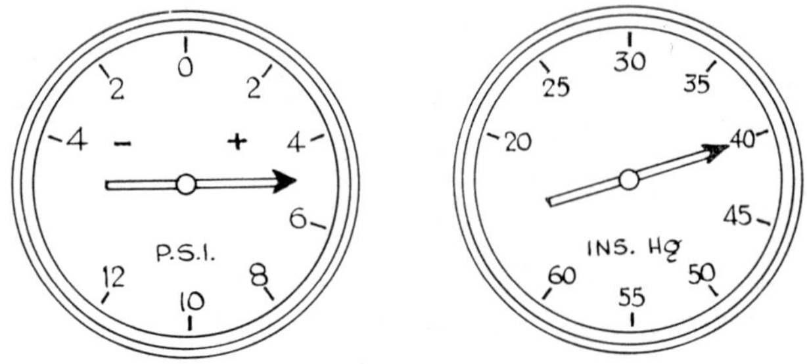

(a) (b)

FIG. 3.16.

To increase the power of an engine so as to reach greater heights than can be attained by the normally aspirated type of engine, it has to be ensured that the pressure on the inlet side of the manifold is not reduced with atmosphere. This is done by boosting the pressure on the engine, or *supercharging*, the supercharger being a means of taking the fuel-air mixture from the carburettor and increasing the pressure of the mixture before passing it to the cylinders. Thus a knowledge of engine power is obtained by measuring this boost pressure.

It was stated on page 40, that a pressure gauge normally measures pressure relative to atmospheric pressure. Thus at sea level with a reading of 5 p.s.i. on the gauge, the manifold pressure would be 5 p.s.i. + atmospheric pressure = 20 p.s.i. approximately. At 20,000 ft., atmospheric pressure is about 7 p.s.i. and therefore a reading of 5 p.s.i. on the gauge would mean a manifold pressure of 5 p.s.i. + atmospheric pressure = 12 p.s.i. This reduced pressure in the manifold must accordingly reduce the engine power, but the ordinary pressure gauge shows no change of reading. Thus an ordinary type of pressure gauge gives no indication of the power of the engine. In order to have some indication of this power it is necessary to have a pressure gauge which reads absolute pressure; i.e. (as in the previous example) a pressure gauge which would read 20 p.s.i. at sea level and 12 p.s.i. at 20,000 ft. thereby showing the operating condition of the engine. The boost gauge, then, can be described as a pressure gauge which measures absolute pressure, although the indication given may be relative to atmospheric pressure. There are two methods of presenting the information given by the boost gauge (Fig. 3.16).

Fig. 3.16(*a*) shows an instrument dial calibrated in terms of p.s.i. and showing a reading of +5 p.s.i. This really means a pressure of 5 p.s.i. above the normal pressure of 15 p.s.i. (approximately), or an absolute pressure of 20 p.s.i. In Fig. 3.16(*b*) the dial is calibrated in inches Hg and is showing absolute pressure directly. In this case the pressure shown is 40 in. Hg and since normal atmospheric pressure is 30 in. Hg then the instrument is showing 10 in. Hg above normal. Remembering that 2 in. Hg = 1 p.s.i. we see that this second gauge is reading 5 p.s.i. above normal; i.e. a reading of +5 p.s.i. on one dial is equivalent to a reading of 40 in. Hg on the other dial.

Owing to the fact that the gauge measures absolute pressure, it is important to consider the reading of the instrument when the engine to which it is attached is not running. At first sight it would appear that in Fig. 3.16(*a*) under these conditions the indicator would read zero, while in (*b*) it would read 30 in. Hg. It will, in fact, still read supercharger pressure which, with the engine not running, is at atmospheric pressure. Since atmospheric pressure is seldom at its nominal value of 14·71 p.s.i. the pointer of the boost gauge will rest on the dial at a point which indicates the actual value of atmospheric pressure at that moment and not necessarily zero or 30 in. Hg.

Gauge Construction. One type of gauge consists of an evacuated capsule which is mounted in an airtight case. Pressure from the manifold is led to the interior of the case by means of a union at the back of the instrument. This pressure, acting on the capsule, operates a link, quadrant and pinion to move a pointer over a scale graduated in p.s.i. or in. Hg. The method of calibration in this instance is the same as for a simple pressure gauge. The drawback of this instrument lies in the fact that the case must be completely airtight to effect accurate readings. A small leak in the case, such as sometimes occurs round the gland of the instrument, may cause under-reading to a considerable degree.

In the bellows type of gauge the instrument functions independently of the case, which need not be airtight but which is usually sealed to exclude dust. Pressure from the manifold is led to a bellows unit which opens out to measure the pressure. A spring is used as the controlling force in the instrument. This arrangement in itself is not, however, sufficient since the total movement of the bellows is the resultant of the forces exerted by the manifold pressure, the spring and atmospheric pressure. It is necessary, therefore, to neutralize the effect of reduced atmospheric pressure with height, so that manifold pressures can be measured directly against the spring. This is done by fitting a second bellows which is identical with the first, except that it is sealed and evacuated.

A pressure entering the lower bellows will cause it to open out to show the change of pressure. With increasing height, however, the reduced value of atmospheric pressure offers less opposition to the movement of the bellows, which would move to show the effect of this reduced opposition. This movement is counteracted by the upper bellows, which tends to open out as in the case of an altimeter. When uniform bellows are used in the upper and lower positions, the forces on each are equal and therefore cancel out, so that movement of the lower bellows takes place only when there is a change in the manifold pressure.

The mechanism employed in the bellows type of boost gauge is similar to that of a simple pressure gauge in that it has a link, quadrant and pinion. There is a slight difference, however, in that a long arm is used to convey movement of the bellows centre-piece to the link operating the quadrant (Fig. 3.17).

Calibration. The boost gauge is calibrated by comparing its readings with that of a mercury manometer when subjected to various pressures or depressions. Adjustments are made to the mechanism to counter any faulty readings, until the instrument lies within the permitted tolerance. The tolerances vary with the range of the instrument and it is essential to determine the tolerance for a particular instrument before proceeding with the calibration.

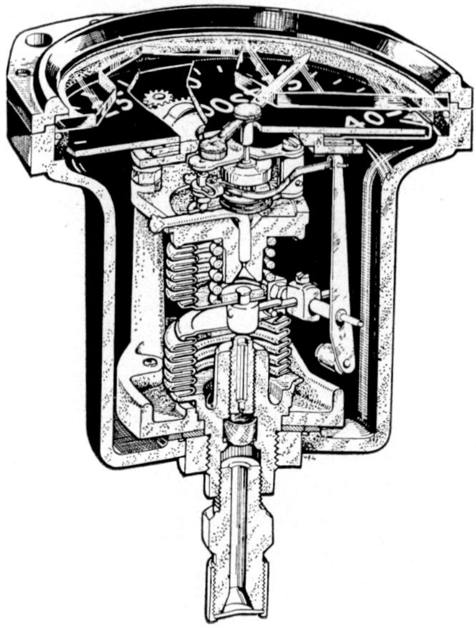
FIG. 3.17. *Boost-type bellows unit.*

There are three adjustments on the bellows-type boost gauges. They are as follows. (*a*) Lever-length adjustment for rough ranging of the instrument. This is effected by pinning the link to the tail of the quadrant at a point nearer to, or farther from, the pivoting point of the quadrant (Fig. 3.17). (*b*) Lever-length adjustment for sensitive ranging of the instrument. This consists of a lever which is clamped to the rocking shaft and which is capable of being moved in or out. (*c*) Lever-angle adjustment for initial setting of the instrument. The correct setting is achieved by pinning the rocking shaft arm to the link by one of several holes in the link.

The instrument is first checked at various points throughout the scale and the errors noted. If the readings at the top and bottom ends of the scale are high, it suggests that the lever angle is not correct. The lever angle is adjusted by pinning the quadrant to the link at a point *farther from the end of the link* in the case quoted.

If readings are high at the top end of the scale and low at the bottom end of the scale, it suggests a lever-length adjustment. Rough adjustment is made by re-pinning the link to the quadrant at a point farther from the quadrant pivot. Final adjustment may be made by loosening the clamping screw on the rocking shaft and moving the bent arm *away* from the rocking shaft to reduce still further the magnification of the pointer movement.

Maintenance. The maintenance required on the bellows-type boost gauge is much the same as for the direct-type pressure gauge since the same kind of mechanism is used. The first requirement is to check the over-load stops. An over-load stop is fitted which limits movement of the quadrant in one

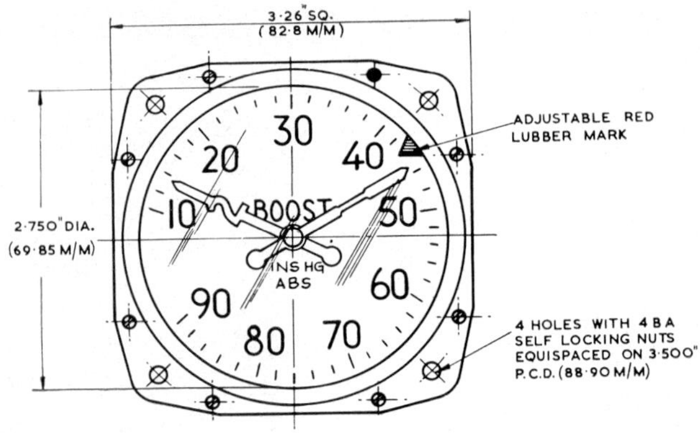

FIG. 3.18.

direction, while an under-load stop limits movement of the bellows in the opposite direction. The under-load stop consists of a long screw, drilled down its centre to provide a restriction to the pressure, which is screwed into the connection at the back of the case (Fig. 3.17). A filter fitted after the restrictor prevents the fine-bore hole from becoming stopped up. The stops should be in such a position that the pointer of the instrument travels *almost* to the centre of the unused portion of the dial on over-load and under-load pressures.

Check that the lubber mark is set to the correct value for the particular aircraft on which the boost gauge is to be used. This lubber mark consists of a red pointer which can be set to indicate the maximum permissible boost allowed in level flight for a given engine. The lubber mark may be attached to a metal ring which is held in place by the glass front of the instrument, or it may be geared to a setting screw which is accessible from the rear of the instrument. Also, make certain that there is no leak in the bellows of the instrument.

Double-boost Gauge. It is possible to have a boost gauge which is capable of indicating the boost pressures of two engines simultaneously. An instrument of this kind, however, merely contains two completely separate gauges with pointers which are concentric and which indicate on a common scale. For identification purposes the pointers may be numbered or marked in such a way as to show to which engine they are connected (Fig. 3.18). Such an instrument may be treated as two separate instruments for maintenance purposes.

CHAPTER 4

Measurement of Airspeed

MEASUREMENT of airspeed is yet another requirement which is carried out by means of pressure measurement. In this instance the pressure measured is a differential pressure related to aircraft velocity. There are two main reasons for requiring to know the speed of an aircraft. (1) It is useful, for navigational purposes, for the navigator to know the speed of an aircraft relative to the ground. (2) It is necessary to know the speed of an aircraft relative to the air surrounding the aircraft, as this airspeed gives an indication of the stability of the aircraft.

An aircraft flies because of the lift force resulting from an airflow over the wings and fuselage of the aircraft produced by the thrust from the engine. The value of lift depends on air density, wing area and wing incidence (angle of wing to airflow). Given fixed dimensions it can be said that lift varies with the square of the velocity of the airflow. That is, lift varies as V^2 varies (where V is the velocity). Airspeed is therefore important as affecting the aircraft safety and control by the pilot.

True Airspeed. In meteorology it is customary to measure airspeed by means of a cup type of anemometer. This is a device which has small cups mounted at the end of arms radiating from a vertical spindle. An airflow past an anemometer of this type causes the cups to rotate the spindle at a speed dependent on the airflow and therefore to givc an indication of the speed of the airflow. Since this anemometer functions independently of air density it can be said to show "true airspeed", slight friction in the mechanism being ignored. Because reduction of air density causes reduction of lift it will be appreciated that at lower densities an aircraft's speed must be increased to maintain a constant value of lift. The cup anemometer, therefore, while indicating true airspeed, would not make allowance for the change of lift on the aircraft.

An airspeed indicator, which depends on air density (among other things) for its operation, will show a reduction of speed with reduced density and therefore "lift" will remain constant with constant indicated airspeed.

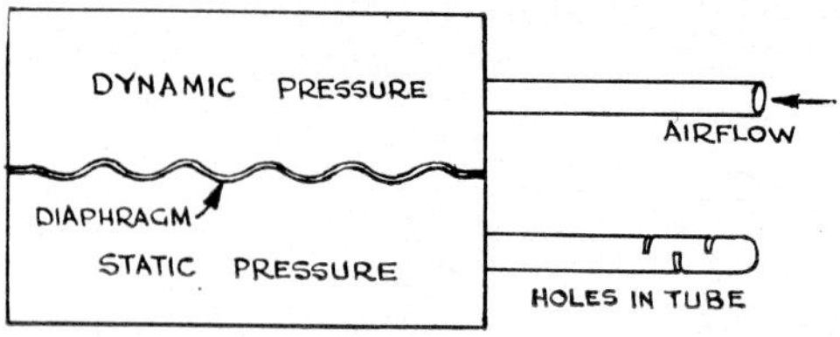

FIG. 4.1.

An indicator of this type is called a tube anemometer. The tube anemometer consists of a sensitive differential pressure gauge which measures the difference of pressure in two tubes. One tube contains air at a dynamic pressure due to the forward speed of the aircraft while the other contains air at the static atmospheric pressure. The tubes are usually combined in a device known as a pressure-head.

Pressure-head. The pressure static system (pressure-head) is sometimes called a pitot tube after the Frenchman Pitot, who used the system for measuring the speed of flow of water in rivers. The pressure head consists of two tubes facing into the flow of air past the aircraft, one tube being open at the forward end, and the other tube closed but having holes or slots cut in its periphery, as shown in Fig. 4.1. Air entering the pressure tube will be trapped in the top compartment of the system and will build up a pressure. The air in the lower compartment, however, will be still air (entering through the slots) at a pressure equal to the pressure of the atmosphere surrounding the static tube. Thus, a differential pressure is built up which gives a measure of the velocity of the airflow past the system.

Bernoulli's theorem for a fluid in motion states that:
Total energy = position energy + kinetic (velocity) energy + pressure energy.

In the tube anemometer, position energy is neglected and the total energy for a unit volume of fluid of density ρ at a static pressure p and moving with a velocity v equals $\frac{1}{2}\rho v^2 + p$. Thus the total energy, the static pressure and the density must be measured to evaluate v.

By reducing the velocity of the medium to zero (as in the pitot/static system of airspeed) the pressure energy equals the total energy; i.e. pressure energy is equal to $\frac{1}{2}\rho v^2 + p$. However, the static tube is measuring p and therefore:

$$p_0 - p = \tfrac{1}{2}\rho v^2$$

where p_0 is the pressure energy, and p is the static pressure.

In effect then, the difference of pressure in the pressure/static system is equal to $\frac{1}{2}\rho v^2$.

For the purposes of this formula it has been assumed that air is a "perfect fluid"; however, tests on a "whirling arm" have shown that it is true for air to an accuracy of 0·1 per cent in the speed range, 1,200–3,000 ft./min.

When considering the compressibility effect of the air in the pitot tube, the difference of pressure, $p_0 - p$, is:

$$p_0 - p = \tfrac{1}{2}v^2\{1 + \tfrac{1}{4}(v/a)^2 + \tfrac{1}{40}(v/a)^2 + \dots\}$$

where a is the velocity of sound in air. Obviously $(v/a)^2$ is a very small quantity, and for practical purposes only the first term is considered, thus:

$$p_0 - p = \tfrac{1}{2}\rho v^2\{1 + \tfrac{1}{4}(v/a)^2\}$$

For calibration purposes the differential pressure $p_0 - p$, or P, is expressed in millimetres of water and, having considered the relative densities of air and water, the formula is expressed as:

$$P = 0{\cdot}012504\ V^2\ (1 + 0{\cdot}43\ V^2 \times 10^{-6})$$

where P is given in mm. H_2O at 15 deg. C., and V is in miles per hour. When V is given in knots the expression becomes:

$$P = 0{\cdot}01658\ V^2\ (1 + 0{\cdot}57\ V^2 \times 10^{-6})$$

(See the Table on page 227.)

The *venturi tube* is another means of creating a differential pressure proportional to airspeed. However, while this tube produces a greater pressure than the pitot/static tube for a given airflow, it suffers from the disadvantage of being less accurate. Since sensitive and accurate differential pressure gauges are available, the low pressure of the pitot/static tube is no drawback.

TYPES OF PITOT TUBE

The simplest type of pitot tube or pressure head is the one previously mentioned where two parallel tubes face into the airflow. Precautions are taken to prevent dirt and water from entering the open-end tube by fitting traps of varying construction. The more usual type of pressure head is the one where the pressure tube is built inside the static tube. A heater placed between the static and pressure tubes prevents icing-up of the tube under adverse weather conditions. Drain holes in the head allow moisture to escape, and these should be kept free of obstruction. These pressure heads are made for a variety of fixings; e.g. under-wing fixing, over-wing, front-of-wing, and so on, but in principle all operate in the same way. Care should be taken to see that the drain hole in the outer (static) tube is on the underside of the pressure head (Fig. 4.2).

Some installations make use of a pressure head with only one connection (pressure), the static pressure being obtained from a *static vent*. This consists of a flat plate which is mounted on the side of the fuselage at a point where there is least disturbance of airflow. A hole in the plate allows static pressure to be led to the airspeed indicator. The result is that a slightly more accurate reading is obtained than that given by the pressure static head. It is important to see that the

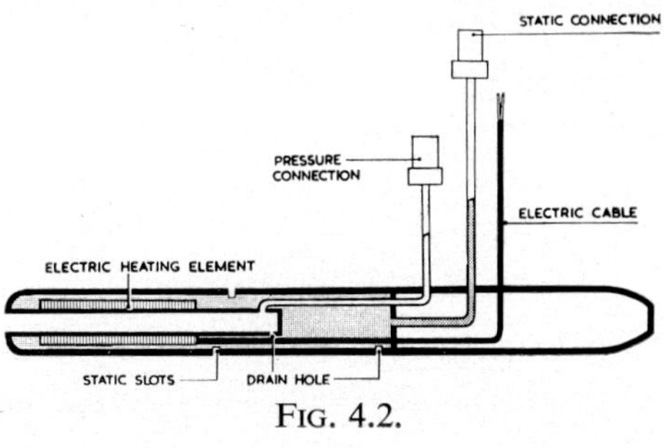

FIG. 4.2.

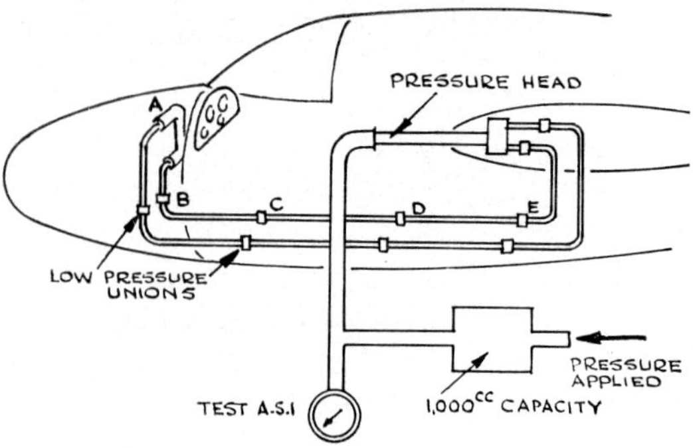

FIG. 4.3.

plate retains its smooth finish and is not painted over.

The maintenance of pressure heads is simple and consists of checking the alignment of the head and keeping it free of foreign matter. The head should be aligned so as to be parallel with the centre line of the aircraft and should be horizontal with the aircraft in the flying attitude.

Testing the Pressure Head. Apart from the usual insulation tests on the heater, the pressure/static system (the head and connecting pipe-lines to the instrument) should be checked for leaks. One method of doing this is shown in Fig. 4.3. The pressure and static pipe-lines have been disconnected from the back of the instrument and interconnected by means of a piece of rubber tubing. An adaptor is then pushed over the pressure head covering the static slots, but not the drain hole, and held in place by means of a clip. The drain hole is sealed with tape. Pressure is now applied to one leg of the T junction on the adaptor, via a chamber housing a capacity of 1,000 c.c. until the airspeed indicator in the other leg reads 150 m.p.h. The pressure tube is now sealed off and any leaks in the system will be shown by a reduced reading in the indicated airspeed. The pointer movement should not decrease at a greater rate than 15 miles in 50 seconds.

To ensure that the drain holes in the static and pressure pipe-lines are free from obstruction, pinch off the rubber tube at *A* (Fig. 4.3), and remove the tape from the static drain hole. The reading of the airspeed indicator

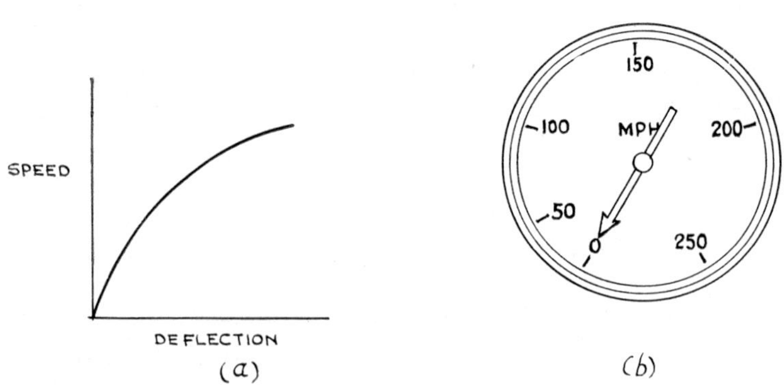

FIG. 4.4.

should now fall off steadily to zero. If the system leaks too much, then the low-pressure unions at *B* are removed; the two pipe-lines are joined by the rubber tube from *A*, and the test is repeated. Further tests may be necessary at *C* and *D*. In this way the fault can be localized and remedied.

Where the static and pressure pipe-lines run together it is an advantage to stagger the unions, as illustrated, so that when the pipe-lines are disconnected there is no possibility of reconnecting them the wrong way round.

Where a static vent is used, tests are carried out separately on the pressure and static lines. The procedure for testing the pressure head, given earlier, can be modified for this purpose. It should be noted that the static line from the vent or pressure head also supplies the altimeter and rate-of-climb indicator with static atmospheric pressure.

AIRSPEED INDICATOR

The airspeed indicator (A.S.I.) is a sensitive differential pressure gauge. The pressure-sensitive element is a capsule mounted in an airtight case, to which is led the pressure pipe-line from the pitot static system. The static connection is led to the interior of the case, which thus contains the low pressure. A difference of pressure will cause the capsule to open out, the movement being proportional to the pressure. A link, quadrant and pinion could be used to convey this movement to a pointer and dial calibrated in terms of m.p.h. or knots. However, the differential pressure varies with the square of the speed and if, therefore, a capsule with a linear pressure/deflection characteristic is used, the speed/deflection characteristic would be as in Fig. 4.4(*a*). This would result in a scale shape that would be uneven, showing larger and larger pointer movements for equal speed changes (Fig. 4.4*b*). To produce a linear scale shape it is necessary to control the characteristic of the capsule and/or the mechanism. Control of the capsule is rather difficult and would involve a characteristic as in Fig. 4.5(*a*), giving a result as shown in Fig. 4.5(*b*). It will be appreciated that very close limits would apply to the capsule as any errors would be magnified by the mechanism conveying movement to the pointer.

Control of the mechanism to produce a linear scale shape can be achieved by changing the lever length as the pointer advances. This can be done in the following way. Movement of the capsule is conveyed to a rocking shaft on which is mounted a long arm. The arm engages in a slot in a toothed wheel, which meshes with a pointer pinion (Fig. 4.6). The lever length is the

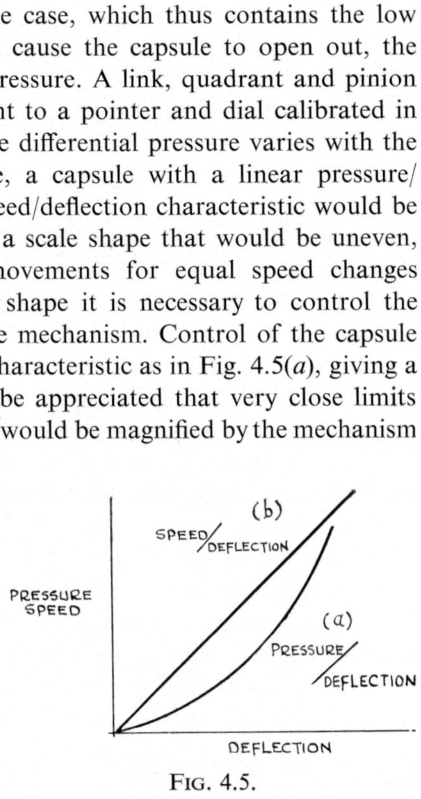

FIG. 4.5.

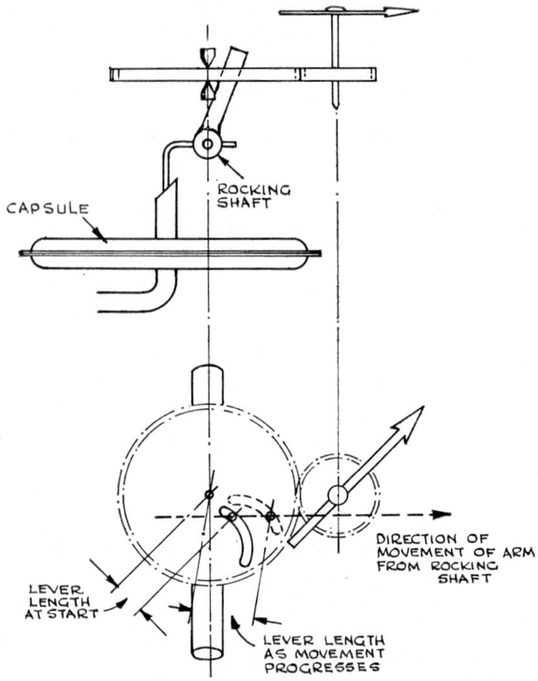

FIG. 4.6.

distance between the pivot of the large wheel and the point of contact between the arm and the slot. As the capsule opens out, the rocking shaft rotates, the arm moves in a straight line and the wheel rotates. The distance between the pivot and arm thus increases with increasing readings, thereby reducing the pointer magnification. It is arranged that this decreasing magnification offsets the increasing rate of capsule deflection, so producing a linear scale shape to the instrument.

In the type of simple airspeed indicator made by Smiths Aircraft Instruments Ltd. a combination of both methods is used, since the capsule characteristic is approaching that desired, while the mechanism enables final adjustment to be made to the scale. The mechanism, however, has been improved so as to reduce the sliding friction which occurs between the arm and the slot. This is done by mounting a roller on the large toothed wheel and fitting a ball-end to the rocking-shaft arm. The effect is to secure a minimum contact area between ball and roller with no sliding friction, since the roller rotates on its axis.

The complete mechanism of the Smith type of instrument is shown in Fig. 4.7, which also illustrates the method of adjusting the instrument. These adjustments are: lever length, which controls the middle range of speeds; axial position of rocking shaft, which controls the top end of the speed range; the lever-angle adjustment to control the lower speed range. Temperature compensation is achieved by means of a bi-metal strip, which conveys movement from the capsule to the rocking shaft.

Maintenance. This can be regarded as being in two parts: maintenance of the instrument as a simple pressure gauge; maintenance of the instrument as part of the pitot/static system. Maintenance of the instrument as a pressure gauge is straightforward and consists of (*a*) checking pivots and bearings for freedom without too much play being evident; (*b*) checking the set of the hair-spring (tension of half turn); (*c*) checking the tension of the U spring which, if too great, will cause difficulty in calibrating the lower end of the scale; (*d*) checking tightness of screws and locknuts. Maintenance of the instrument as part of the installation involves checking the case for leakage.

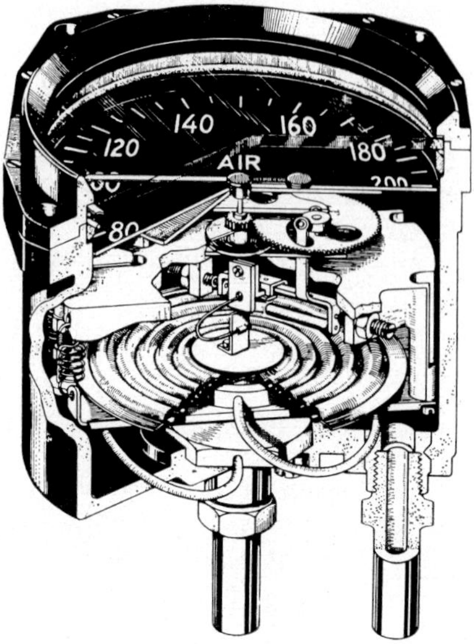

FIG. 4.7. *Simple airspeed indicator.*

The instrument is tested against a standard manometer and the readings compared. The indicator under test should agree with the manometer to within a very small percentage tolerance, depending on the range of the instrument (generally about 1 per cent).

The manometer may be a simple U tube filled with water or mercury and calibrated according to the formula given on page 57, or it may be a little more elaborate with a dial and pointer to show the mercury level. Having checked the calibration in this way, the case is tested for leakage by applying suction to the static connection of the instrument until the pointer reads 150 m.p.h., sealing off the instrument and noting that any pointer movement does not exceed 15 m.p.h. in 10 seconds.

PRESSURIZED INSTRUMENTS

When leak-testing pressurized instruments it is necessary to ensure that the case will withstand an external pressure of 15 lb. per sq. in. The method of testing is as follows. Place the instrument in a pressure chamber capable of withstanding a pressure of one atmosphere. Join the static and pressure connections of the instrument with rubber tubing and a tee-piece, taking one leg of the tubing out of the pressure chamber and connecting it to a small water manometer (Fig. 4.8). Apply a pressure of 15 p.s.i. to the

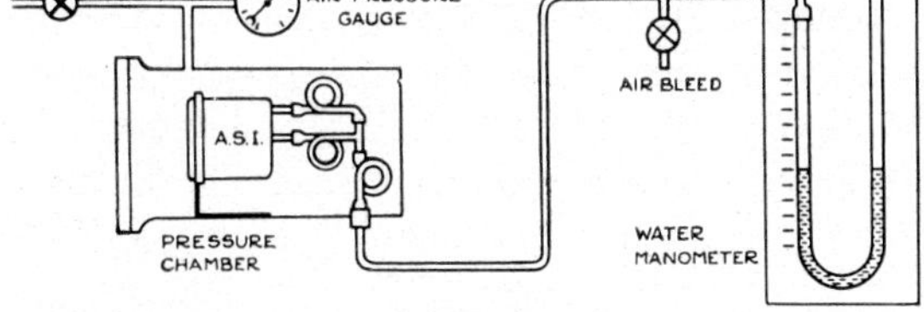

FIG. 4.8.

pressure chamber and watch the U tube for any change of levels, which will thereby denote a leakage in the casing. Any leaks thus shown on the U tube must not exceed 1 c.c. per minute.

Calibration. If, during the test against a master, the instrument shows errors outside the permissible tolerance, it is necessary to remove the errors by adjustment of lever length, lever angle and so on. As previously stated, each part of the range is covered by a particular adjustment and the procedure of calibration of the instrument is outlined in the following paragraphs.

Assume the range of the instrument to be 60 to 400 m.p.h.

(*a*) Connect indicator under test to manometer or master airspeed indicator.

(*b*) Apply pressure until master reads 140 m.p.h. and set pointer on test instrument at this reading.

(*c*) Increase pressure until master reads 300 m.p.h. and note reading of test indicator.

(*d*) If reading low, reduce the lever length; if reading high, increase the lever length (Fig. 4.7).

(*e*) It may be necessary now to repeat (*b*).

(*f*) Having obtained correction between 140–300 m.p.h. increase the pressure further until master reads 400 m.p.h. and note reading of test indicator.

(*g*) Correction is achieved by axial positioning of the rocking shaft. Movement of the shaft to the right decreases the topmost reading.

(*h*) Check range 140–400 m.p.h.

(*i*) Set master at 50 m.p.h. and position pointer to test instrument at this reading.

(*j*) Apply pressure until master reads 140 m.p.h. and note test indicator reading.

(*k*) If too low, adjust lever angle (Fig. 4.7) so as to move the capsule away from the rocking shaft.

(*l*) Check finally through the complete range, making slight adjustments as necessary until indicator reads within the permissible tolerance.

SENSITIVE AIRSPEED INDICATOR

The sensitive airspeed indicator fulfils the same function as the simple airspeed indicator described on page 59. It is calibrated to the same laws and operates from the normal pitot/static system; as its name suggests, it provides a more sensitive indication. This is achieved by an increase in the gear train from the capsule, so that two pointers may be moved over an evenly calibrated dial. Because of the increase in the gear train, it is necessary to provide more power to operate the gears. The additional power comes from the use of a double capsule. This capsule assembly has a linear pressure/deflection

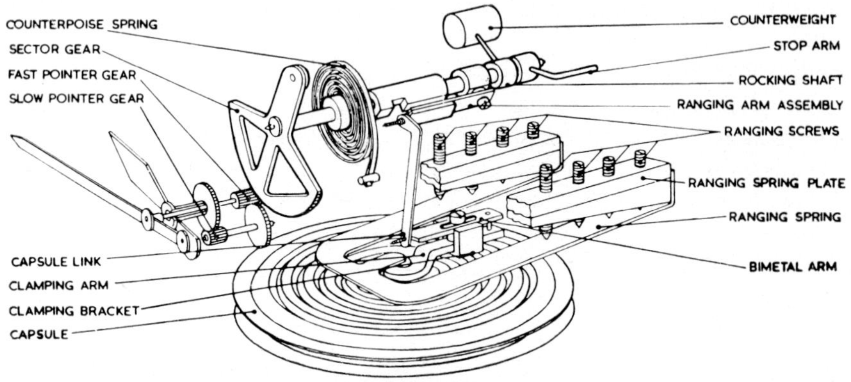

FIG. 4.9.

characteristic which is closely controlled and requires more careful forming than the capsule used in the simple type of airspeed indicator.

The mechanism of the sensitive airspeed indicator differs from that of the simple A.S.I. in the method by which a linear scale is obtained. In the sensitive airspeed indicator the scale shape is governed by varying the rate of a "tuning" spring which bears on the capsule assembly. The spring rate is made to increase as the speed increases, by the suitable positioning of screws against which the spring bears as the capsule lifts. The movement can then be carried via a normal rocking shaft, quadrant and gear train to the two pointers. The complete mechanism of the sensitive airspeed indicator is shown in Fig. 4.9. At lower pressure the tuning spring is not necessary, and it is arranged that the capsule comes into contact with the spring when a speed of 120 m.p.h. is reached. Maintenance and testing are the same as for the simple A.S.I. previously described.

Calibration. Calibration of this instrument is effected by means of adjustment of the lever length (ranging arm), lever angle and tuning screws. (If the bi-metal strip has been moved during dismantling it will be necessary to reset this before starting calibration proper.) To set the bi-metal strip in its correct position, the readings of the instrument under temperature conditions are compared with the normal readings. If the readings are high when hot and low when cold, the clamp which locks the bi-metal strip should be moved away from the centre of the capsule.

The instrument can now be ranged or calibrated by connecting it to a master instrument and adjusting, if necessary, as follows. Starting with the capsule boss against the instrument frame:

(*a*) Apply pressure to the instrument until the master reads 60, 80, 100 and 120 m.p.h. (or knots). Adjust on the ranging arm until the instrument is within tolerance.

(*b*) Continue to increase the reading to 140 m.p.h. (or knots) when the instrument under test will begin to show positive errors. Adjust the outermost tuning screws until the instrument is in tolerance at this reading.

(*c*) Readings are now taken at intervals up to 340 m.p.h. and errors counteracted by adjustment to the ranging arm. Moving the arm nearer to the rocking shaft increases the reading.

(*d*) Any adjustment of the arm at this stage will affect the readings at the lower end of the scale, but these can be reduced by moving the capsule assembly away from the frame. (This adjustment is effective only over the lower speed range.)

(*e*) Repeat the test of the instrument between 50–340 m.p.h., making slight adjustments as previously described until the instrument is within tolerance over this range.

(*f*) Further increase of reading will show a tendency on the part of the test instrument to indicate positive errors. These are corrected up to 440 m.p.h. by adjusting on the *fourth* set of tuning screws.

(*g*) Errors beyond 440 m.p.h. are brought inside tolerance by adjusting successive tuning screws.

(*h*) The stop on the rocking shaft is now adjusted so that the instrument reads 44–48 m.p.h. in the static position.

(*i*) The upper limit of approximately 5 m.p.h. over the maximum reading is effected by means of two overload stop screws at the centre of the tuning-spring bridge piece.

MAXIMUM SAFE AIRSPEED INDICATOR

With high-speed aircraft additional problems are encountered as the speed of sound is approached. At speeds nearing the speed of sound the aerodynamic stability of the aircraft changes, due to the breaking up of the airflow over the aerofoils. The effect of this can be so great as to make the aircraft extremely difficult, if not impossible, to control, even if the airframe is capable of withstanding the severe strain imposed on it by these conditions.

This critical speed will vary according to the type of aircraft and with altitude, because of the change of air density which affects the aerodynamic characteristics of the aircraft. Because of this density effect, the speed of sound, too, will vary with altitude, but it can be shown that if the ratio of aircraft speed to sound speed is kept constant the aerodynamic stability of an aeroplane remains unchanged. This ratio, aircraft speed/speed of sound, is known as the Mach ratio and is given a number according to the type of aircraft; e.g. an aircraft whose stability begins to break down at 570 m.p.h. at sea level is said to have a Mach number of $570/761 = 0{\cdot}75$.

Some aircraft carry an instrument known as a machmeter to indicate to the pilot the Mach number of the aircraft at any given instant. The machmeter is carried in addition to the normal airspeed indicator so that the pilot needs to check two readings: (1) the indicated airspeed; (2) speed relative to the speed of sound. In the Maximum Safe Airspeed Indicator (Smiths) both these readings, in effect, are given on one dial.

The principle of the maximum safe airspeed indicator will be clearly understood if we take as an example an aircraft with a Mach number of unity. In fact, this means that the maximum true airspeed of the aircraft is equal to the speed of sound (761 m.p.h. at sea level).

Imagine an airspeed indicator which reads true airspeed; then it would be sufficient to place a red mark on the dial at 761 m.p.h. and to warn the pilot not to exceed this reading at sea level. However, increasing altitude brings about a reduction of air density and a corresponding decrease in the speed of sound; therefore, the red warning mark would have to be moved to a lower reading as the aircraft climbed. In practice the red mark is a pointer which indicates an airspeed equivalent to a given Mach number. The pointer is attached to an altimeter capsule, which causes it to move backwards over the dial as altitude increases.

To obtain true maximum speed and true airspeed it is necessary to correct for temperature and density; however, since this means applying the same correction to both sides of an equation the ratio *indicated airspeed/indicated Mach number* is the same as *true airspeed/true Mach number* and the instrument requires no special compensation. All aircraft do not have the same Mach number, however, so the instrument has a subsidiary scale which can be set to the Mach number of any given aircraft. This setting fixes the position of the maximum safe airspeed pointer to one where the indicated airspeed is equivalent to the appropriate Mach number. There are then, two separate and independent movements in one case; a normal airspeed indicator movement and an altimeter movement which reads in terms of *indicated* airspeed.

Airspeed Movement. This consists of a brazed beryllium-copper capsule operating a pointer through a magnifying mechanism. The capsule characteristic is controlled so as to give a linear deflection over the lower portion of the airspeed scale, which is quite open. At 150 knots a restraining spring reduces the rate of the capsule and therefore the scale is closed up, but still linear. Calibration is in two parts.

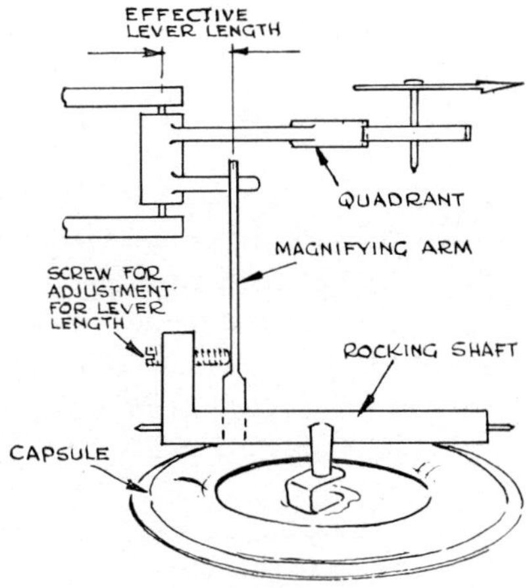

FIG. 4.10.

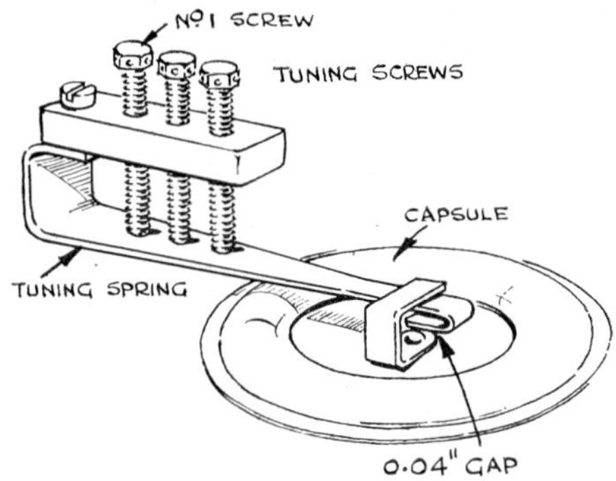

FIG. 4.11.

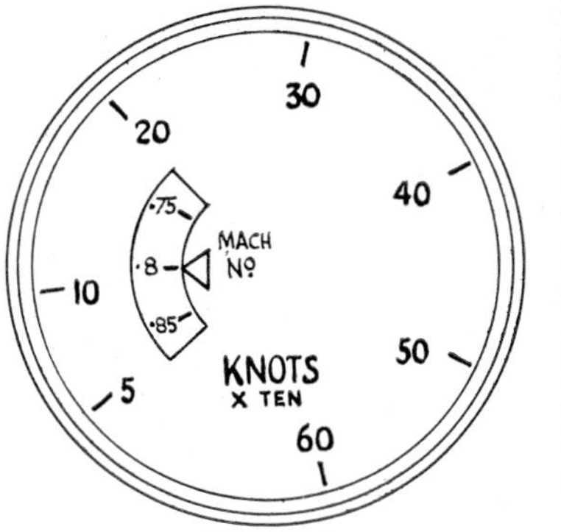

FIG. 4.12.

(*a*) Calibration of the open part of the scale. This is achieved by varying the lever-length mechanism so as to increase or decrease the magnification rate. The adjustment is made by a screw which bears on the arm (Fig. 4.10).

(*b*) Calibration of the remainder of the scale. This is effected by means of the tuning screws (Fig. 4.11) which vary the rate of the spring operating on the capsule. After the initial calibration (as in *a*), pressure is continually applied while a check is made of the readings of master calibrator and indicator under test. The indicator will soon begin to show positive errors and when these errors are outside the permitted tolerance the first (outer) tuning screw is screwed down on to the spring controlling the capsule. The reading of the instrument having been corrected in this way, pressure is once more applied until again the instrument shows positive errors outside the permitted tolerance, when the second screw is brought into play. This procedure is continued, using successive screws, until all the airspeed readings lie within this permitted tolerance. (It should be noted that there is an initial gap of something like 0·04 in. between capsule and tuning spring, which allows free movement of the capsule over the lower or open part of the scale. At this stage the spring should be parallel with the back plate.)

Altimeter Movement. This consists of an evacuated capsule assembly which drives the maximum safe airspeed pointer through a special linkage, to indicate an airspeed equivalent to a given Mach number at any particular altitude. Provision is made in the linkage for manual adjustment of the Mach number of the instrument from 0·5 to 1·0. The Mach number thus set is shown on a subsidiary dial as in Fig. 4.12.

Since some aircraft have an overriding maximum speed at low altitude, provision is made for an adjustable stop to be fitted. This controls the maximum position of the maximum safe airspeed pointer; e.g. supposing an aircraft has a Mach number of 0·75, this is equivalent to a speed of $0{\cdot}75 \times 761$, or 571 m.p.h. (at sea level). However, an overriding maximum speed of, say,

550 m.p.h. may be imposed on the aircraft at this height. These conditions are provided for by adjusting a stop so that the maximum safe airspeed equivalent to a given Mach number is shown by the maximum safe airspeed pointer until that speed reaches 550 m.p.h., whereupon this pointer is "left behind" at this value and the mechanism moves on. With increasing altitude, the mechanism once more picks up the maximum safe airspeed pointer and moves it backwards to show a reading appropriate to the Mach number.

The altimeter movement is shown in bold outline in Fig. 4.13, where the action of the special linkage and the overriding maximum stop can be seen. (The reason for the special linkage is discussed later.) As the capsule moves down with decreasing height, the curved arm will press down on the roller, so moving the linkage and the attached quadrant to operate the maximum safe airspeed pointer. When the arm from the quadrant comes into contact with the overriding maximum stop, the pointer becomes fixed, but a spring in the special linkage allows the curved arm and linkage to continue to move. In practice the spring device in the special linkage is much superior to the schematic arrangement shown in the inset to Fig. 4.13.

Calibration. The calibration of the altimeter movement is carried out with the instrument showing a fixed Mach number. The number chosen for this initial calibration is 0·8. There are three essential conditions to be observed before any adjustments are made to the altimeter mechanism:

(*a*) The roller on the special linkage must contact the curved arm at such a point that the distance between the centre of the roller and the centre of the altimeter rocking shaft is 0·6 in.

(*b*) The Mach dial must be at 0·8 exactly.

(*c*) The maximum safe airspeed pointer must be at a reading equivalent to a Mach number of 0·8 subject to temperature and density corrections being applied; e.g. under standard conditions of temperature and density the reading equivalent to 0·8 is 609 m.p.h. (528 knots).

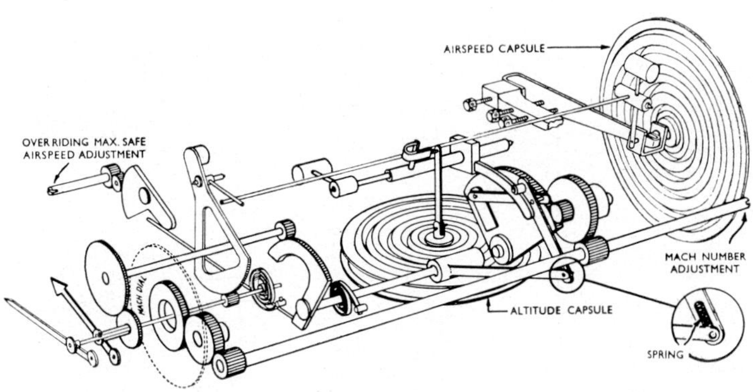

FIG. 4.13.

Table 4.1

CALIBRATION OF ALTIMETER MECHANISM

Height in feet	*Pressure (millibars)*	*Scale*	*Maximum safe airspeed pointer readings at given Mach number settings*					
			0·5	0·6	0·7	0·8	0·9	1·0
0	1013·2	M.P.H.	380	457	533	609	685	761
		Knots	330	396	462	528	594	660
10,000	696·9	M.P.H.	319	383	450	517	586	656
		Knots	277	332	391	449	509	569
20,000	453·3	M.P.H.	262	317	373	430	489	550
		Knots	227	275	324	374	425	478
30,000	300·8	M.P.H.	212	257	303	350	400	451
		Knots	184	223	263	304	347	391
40,000	187·4	M.P.H.	168	203	240	279	319	360
		Knots	146	177	209	242	277	313
50,000	115·9	M.P.H.	132	160	190	220	252	286
		Knots	115	139	165	191	219	248

Having observed these conditions the movement is placed in a vacuum chamber and pressure is decreased. Readings are taken from a barometer (as indication of height) and the instrument under test. These readings should agree with the figures in Table 4.1.

If the instrument is under-reading it is necessary to move the calibrating arm on the altimeter rocking shaft. Moving the arm into the shaft increases the range of the instrument and moving it out of the shaft reduces the rang$\cdot$.

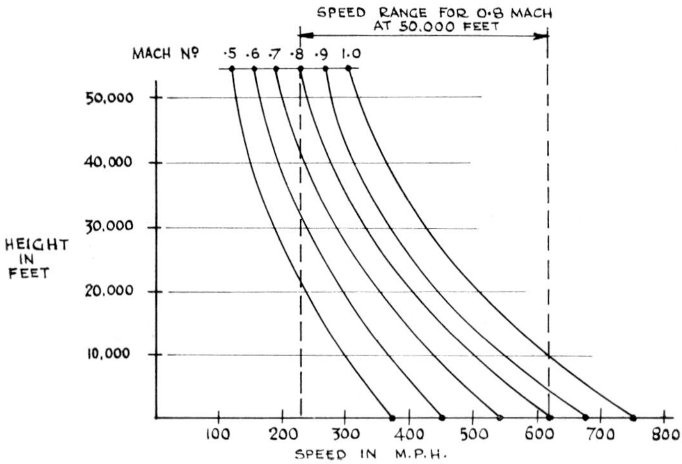

FIG. 4.14.

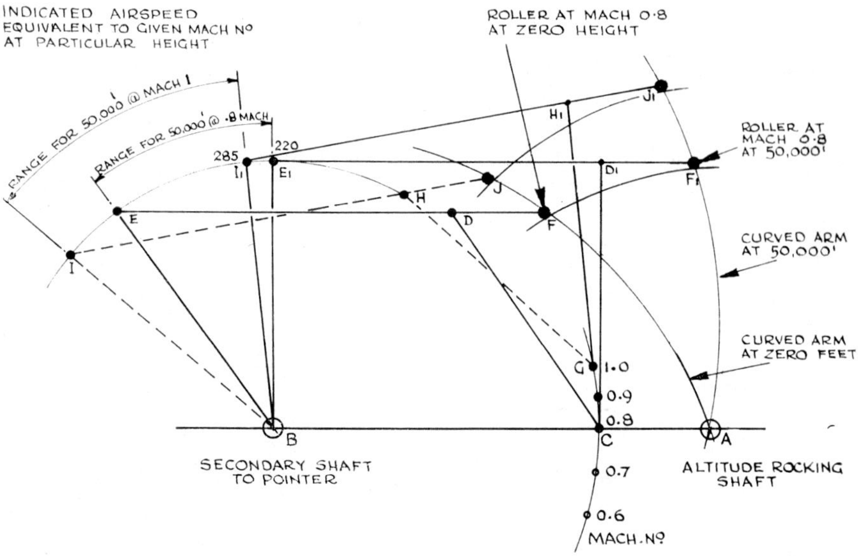

FIG. 4.15.

Small changes in calibration can be made by moving the capsule up or down; a screw in the capsule plate permits this adjustment.

Having calibrated the altimeter mechanism to the figures shown in Table 4.1 (at a Mach dial setting of 0·8), it is necessary to check that manual adjustment of the Mach dial to values other than 0·8 gives the required maximum safe airspeed pointer reading; e.g. if the readings of the pointer at 0·7 and 0·9 were closer to the 0·8 readings than the Table indicates, it is necessary to correct by loosening the curved arm on the altimeter rocking shaft and moving it downwards. Having replaced the pointer at the correct reading, manual adjustment of the Mach dial is again carried out and further correction made, if necessary, until the pointer readings agree with those shown in the Table at each Mach number setting.

Special Linkage. The reason for the special linkage can now be considered, and possibly a graph of speed/height for a given Mach number may help in this consideration (Fig. 4.14). The speed/height characteristic is not linear and whereas at 0·8 Mach the speed range for 50,000 ft. is 389 m.p.h., the range for a similar height for 0·7 and 0·9 Mach is 343 and 437 respectively. This means that the speed range for a given change of height must be varied according to the Mach number, and it is the special linkage already mentioned which provides for this variation.

The functioning of this linkage is shown in Fig. 4.15. The linkage *BCDE*, carrying the roller *F* bearing against the curved arm, is so arranged

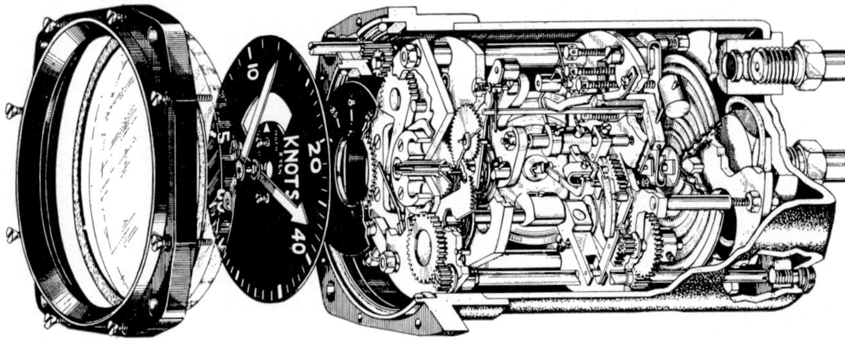

FIG. 4.16. *Maximum Safe Airspeed Indicator.*

that at a Mach number of 0·8 the side *BC* of the parallelogram formed by the linkage coincides with an imaginary line (*A* — *B*) passing through the centres of the rocking shaft and the secondary shaft. These conditions are achieved when the distance between centres from the rocking shaft to the roller is 0·6 in. At 0·8 the curved arm, in moving from zero to 50,000 ft. allows the linkage to move so that the indicated airspeed change is from 609 to 220 m.p.h. (389 m.p.h.) when the position of the linkage is BCD_1E_1 with the roller at F_1.

When the Mach number dial is manually adjusted to Mach 1·0 the position of the linkage is changed so that the indicated airspeed of the maximum safe airspeed pointer is 761 m.p.h. and, while a parallelogram is still formed, no one side coincides with the line *A* — *B*. The linkage now is in position *BGHI* with the roller at *J* and when the curved arm moves from zero to 50,000 ft. the roller moves to J_1 and the linkage to BGH_1I_1. It will be seen from Fig. 4.15 that this movement is greater, for the altitude change, than the movement of the linkage at 0·8 Mach, and so caters for the range of 475 m.p.h. required at Mach 1.0. Similarly, if the Mach number dial is manually set to 0·7, for instance, the linkage is so changed that the speed range from 50,000 ft. is reduced to 343 m.p.h.

The adjustment for the overriding maximum stop is found at the bottom left-hand corner of the instrument in a position usually reserved for fixing the instrument to the panel. The complete mechanism is shown in Fig. 4.16.

MACHMETER

In indicating the Mach number of an aircraft the machmeter must respond to changes of airspeed and changes of the speed of sound. In the Kelvin instrument both these changes are shown by a common pointer to indicate the Mach ratio at any instant. The change in airspeed is registered in the conventional manner by means of a capsule which measures the difference in static and dynamic pressures in the pitot system. Any change in the speed of sound will occur only because of a change in the density of the medium through which the sound is passing. Therefore, an altimeter movement can be used to denote changes in the speed of sound with altitude.

The *airspeed movement* comprises a capsule which is controlled by a variable spring suitably adjusted to ensure the required characteristic. The movement of the capsule is conveyed via a push rod to a rocking shaft which drives a sector and pinion by means of a lever arm (Fig. 4.17). Thus the indication could be of a ratio of the measured airspeed to the speed of sound, but *only* for a fixed speed of sound; e.g. suppose the measured airspeed was 570 m.p.h. then the ratio of this speed to the speed of sound is $570/761 = 0{\cdot}75$.

Now consider the same measured airspeed at an altitude of, say, 10,000 ft., then the ratio will have changed although the speed is the same; e.g. the ratio is $570/650 = 0{\cdot}87$, since the indicated speed of sound at 10,000 ft. is approximately 650 m.p.h. With a constant airspeed, therefore, an increasing reading must be obtained from the instrument pointer, and this increase must be proportional to increased altitude. An altimeter movement is therefore used to provide the change of indication with changing altitude.

The *altimeter movement* comprises an evacuated capsule which responds to changes of barometric pressure and which operates so as to vary the magnification rate of the airspeed movement. The variation is achieved by arranging that the capsule moves the airspeed rocking shaft along its axis, thereby changing the point of contact of the operating pin with the lever arm to vary the lever length. Note that an increase in altitude will cause the capsule to open out, so moving the rocking shaft to reduce the lever length and accordingly increase the Mach reading as required (Fig. 4.18).

Testing, Maintenance and Calibration. Since the machmeter comprises

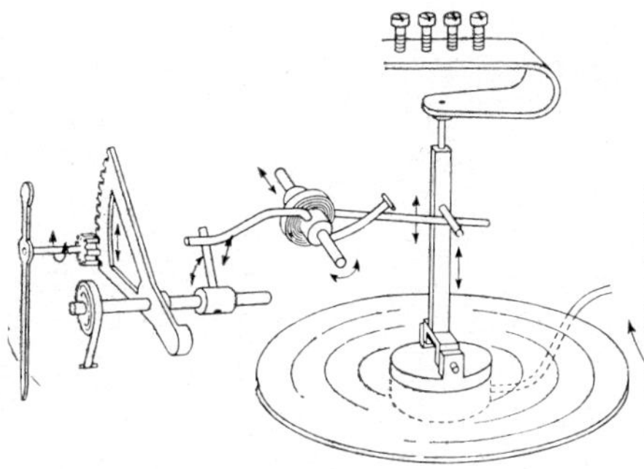

FIG. 4.17.

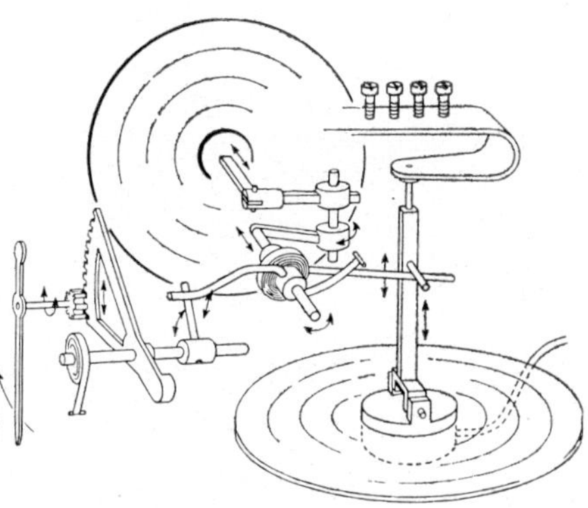

FIG. 4.18.

Table 4.2

MACH NUMBER, AIRSPEED AND ALTITUDE

	A.S.I. reading in knots corresponding to a given altitude in I.C.A.N. ft.						
Mach No.	0	1,000	2,000	10,000	20,000	30,000	40,000
0·5	330	324	319	277	228	184	146
0·6	396	390	383	333	275	223	177
0·7	462	455	448	391	324	263	209
0·8	528	520	512	449	374	304	242
0·9	594	586	577	509	425	347	277
1·0	—	—	—	569	478	391	313

both airspeed indicator and altimeter, the equipment normally associated with those instruments is required for test purposes. The machmeter is placed in a test chamber with its static connection open to the chamber but its pressure connection brought out to an airspeed calibrator. A barometer is connected to the test chamber and the pressure in the chamber is adjusted to equal the standard sea-level pressure. This pressure is held while the airspeed pressure is changed to produce changes in Mach numbers on the instrument dial. The process is repeated using static pressures in the chamber corresponding to altitudes other than zero. Table 4.2 relates Mach number, airspeed and altitude.

The maintenance and calibration of machmeters and maximum safe airspeed indicators is more difficult than for the conventional airspeed indicator. The relevant instructions are detailed and lengthy, and reference should always be made to manufacturers' literature.

CHAPTER 5

Measurement of Height

To measure height in everyday life is usually a simple matter of placing a scale, marked in feet and inches, against the object to be measured. To measure aircraft height above the ground a practical indicator is required that will show, instantly, height and/or changes of height to the pilot and navigator.

The introduction to Chapter 3 outlined a laboratory experiment which determined pressure by measuring the height of a column of liquid. Conversely, the height (*h*) of a column of liquid of density (*d*) can be determined by measuring the pressure which causes the change in height. In the previous example:

$$h = 10{\cdot}2 \text{ in. of mercury (Hg);}$$
$$d = 0{\cdot}49 \text{ lb. per cu. in.;}$$
$$\text{and } p = hd = 10{\cdot}2 \times 0{\cdot}49 = 5 \text{ lb. per sq. in. (p.s.i.).}$$

Similarly, if $P = hd$, then $h = P/d$. Thus, knowing the pressure, the height can be determined: for example, if $P = 6$ p.s.i., $h = 6/0{\cdot}49 = 12{\cdot}24$ in. This method can be used for measuring the height of an aircraft above the earth's surface, because between the two is an atmosphere having weight and density.

Consider a unit area on the earth's surface. There is, acting on each unit area a column of air, say *AB*. The height, *h*, of *AB* is given by P/d, where *P* is the pressure of the atmosphere and *d* is the air density. Considering an aircraft at an altitude *x* (Fig. 5.1), then because the column of air above it is less than that on the earth's surface the pressure is also less; that is, $Ax = P_x/d$, where P_x is the atmospheric pressure at *x*, and *d* is the air density. Now the height of the aircraft above the ground is $xB = AB - Ax$ as indicated in Fig. 5.1.

$$\text{but } AB = P/d, \text{ and } Ax = P_x/d, \text{ therefore}$$
$$xB = AB - Ax = P/d - P_x/d$$

and the height, *x*, of the aircraft above ground is

$$\frac{P - P_x}{d}$$

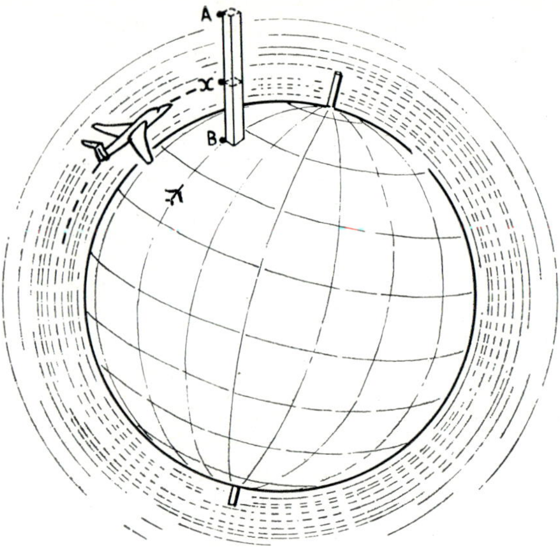

FIG. 5.1.

where P is the atmospheric pressure on the ground, P_x is the atmospheric pressure at x, and d is the air density.

It will be noted that it is the pressure difference between two points which indicates the height between those points and that while a U tube can be used for pressure measurement, it is effective only in regard to pressure *relative* to atmosphere and not to the pressure of the atmosphere itself. For this purpose another type of mercury column can be used. This consists of a single glass tube, sealed at one end, completely filled with mercury. The filled tube is temporarily sealed at the open end, and is then inverted and dipped into a bowl of mercury, and the temporary seal removed. The weight of the column will cause the mercury to start to run out of the tube into the bowl and this will leave a vacuum at the top of the tube. At some point the weight of the column of mercury left in the tube will balance the pressure of the atmosphere acting on the surface of the mercury in the bowl (Fig. 5.2).

Since $h = P/d$, and P is 14·7 normally at sea level, d being 0·49 lb. cu. in., then $h = 14{\cdot}7/0{\cdot}49 = 30$ in. This shows that the normal atmospheric pressure on the earth's surface will support a column of mercury 30 in. high.

It is obvious that at some altitude above the ground where the atmospheric pressure is reduced, the height of the column of mercury that can be supported will also be reduced. The difference in levels of the column at two points is therefore a measure of the difference in height between those two points; for example, it has been shown that $h = P - P_x/d$, where P is the pressure at sea level, P_x is the pressure at a given altitude, and d is the air density between

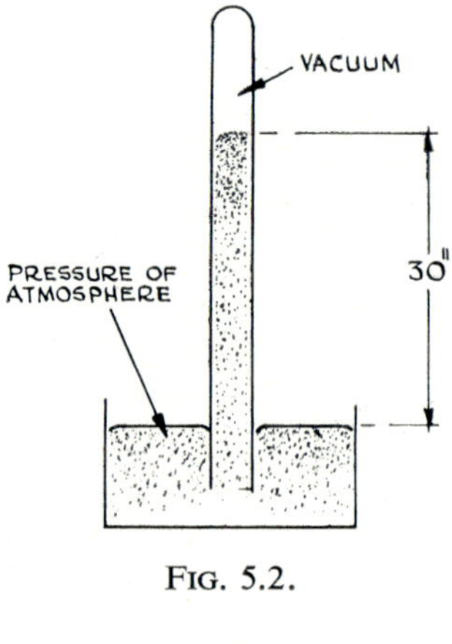

FIG. 5.2.

these two points. Now the pressure P at sea level is 30 in. Hg (normally), and if P_x is, say, 28 in. Hg, then the altitude at x is:

$$h = 30 - 28/d = 2/d \text{ in.}$$

Since d is expressed in lb. per cu. in., the pressure must be expressed in p.s.i., so that 2 in. Hg = 1 p.s.i. Therefore $h = 1/d$, and since d is the air density then:

$$d = \frac{0{\cdot}0012 \times 2{\cdot}54^3}{453{\cdot}6}$$

$$\text{thus } h \text{ in inches} = \frac{1}{\dfrac{0{\cdot}0012 \times 2{\cdot}54^3}{453{\cdot}6}}$$

$$\text{and } h \text{ in feet} = \frac{453{\cdot}6}{0{\cdot}0012 \times 2{\cdot}54^3 \times 12}$$

$$= 1{,}920 \text{ feet approximately.}$$

A mercury column of the kind just discussed is known as a barometer, and the meteorologist's unit of pressure is termed a *bar*, which represents a pressure of 1 megadyne per sq. cm. The bar is a rather large unit, so it is divided into millibars (mb), this sub-unit being equal to 0·001 bar.

The normal atmosphere creates a pressure of 1·0132 bars (1013·2 mb) which is equivalent to 14·7 p.s.i. or 30 in. Hg. A scale calibrated in inches or millimetres can be made and attached to the instrument to provide a means of measuring changes in atmospheric pressure. The problem of measuring height by this means is, however, complicated by the effects of temperature and gravity, which also affect the height of the mercury column. The effect of temperature is such that the specific gravity of mercury is reduced with increasing temperature, thereby increasing the length of the column for the same atmospheric pressure and so giving an apparent change of height. Gravity varies with latitude and height above mean sea level. The effect of latitude is such that the length of the mercury column will increase with a decrease in latitude. This is because the value of gravity varies with latitude, being greater at the poles than the equator. This can be understood if it is remembered that the earth is not spherical but is shaped rather like an orange. If one imagines the force of gravity as acting from the centre of the earth, it has a shorter distance through which to bring about attraction of bodies on the earth's surface at the poles than it has at the equator. At the poles (latitude 90 deg.) the mercury column exerts a greater force and, assuming atmospheric pressure and temperature to be constant, the mercury column will be shorter at the poles than at the equator. Gravity varies with the height above sea level, its value falling with increasing height.

It will be appreciated that a barometer at mean sea level, latitude 65 deg., temperature 15 deg. C., will not give the same reading as a barometer at

5,000 ft., latitude 65 deg., temperature 15 deg. C., even if the atmospheric pressure is the same, and that a correction should be made for the height; this correction is very small (something like 0·003 in. per 1,000 ft.) and in practice is often neglected. In some parts of the world, however, where the height of the measuring station may be 5,000 ft. above sea level, it is important to include this altitude correction.

Standard and London Laboratory Conditions. It has been shown that the barometer will show different readings at various parts of the earth's surface and therefore when taking a reading reference should be made to some known standard. The "standard conditions" to which a barometer is calibrated to give accurate readings are those of mean sea level altitude with temperature at 0 deg. C. and at latitude 45 deg. In Britain, since the standard barometer would have to be corrected for latitude and temperature, a "London Laboratory" barometer is often used. This is calibrated to give accurate readings at mean sea level with a temperature of 16·6 deg. C. and a latitude of 51½ deg.

Whatever the calibration of the barometer, a mercury thermometer is attached in such a way that its bulb is in close proximity to the mercury of the barometer and thus gives an indication of any temperature change from "standard" or "London Laboratory" conditions. Thus, the barometer can be used for measuring height, bearing in mind the formula: height $= (P - P_x)/d$, where $P - P_x$ is the pressure difference and d is the air density and taking into consideration the changing values of g and temperature. (It should be noted that d is not constant and results in a non-linear pressure/height curve as mentioned on page 26.

Measuring Height with a Barometer. It has been shown that the mercury barometer will register a change of pressure with changing height. One could, therefore, install such a barometer in an aircraft and measure off the scale in terms of feet or metres (altitude) and so call the barometer an altimeter. In making such a scale, consideration would have to be given to changes of temperature and gravity and the scale would therefore be calibrated to conform to one of two laws: (1) I.C.A.N.; (2) isothermal.

(1) The I.C.A.N. law states that with a mean sea level pressure of 1013·2 mb and a temperature of 15 deg. C. the temperature falls 1·98 deg. C. per 1,000 ft. up to a height of 36,090 ft., after which it remains constant.

(2) The isothermal law assumes a mean sea level pressure of 1013·2 mb and a temperature of 10 deg. C. which remains constant at all heights.

It is obvious from this that the I.C.A.N. law, while not being absolutely accurate, is certainly of more value than the Isothermal law, and nowadays it is a general rule to calibrate altimeter scales according to the I.C.A.N. law (more recently superseded by the I.C.A.O.). This would necessitate only corrections for change of gravity. However, this can be dismissed since, while the mercury barometer is useful as a standard it is not convenient to use in an aircraft and in its place an aneroid barometer is used which is mass balanced and therefore requires no correction for g.

It may be thought that all the foregoing has been of little consequence, seeing that the mercury barometer is not used in an aircraft to measure height, but it must be emphasized that the mercury barometer is the standard against which the aneroid barometer or altimeter is calibrated and therefore a knowledge of barometry is essential to the person calibrating an altimeter.

An *aneroid barometer* is simply a barometer "without liquid" and must therefore be a device which measures atmospheric pressure changes by mechanical means. The usual aneroid movement consists of two thin metal diaphragms which are soldered together at the edges to form a small flexible chamber or capsule. This capsule is evacuated to a high degree of vacuum so that the tendency is for the capsule to collapse because of the force exerted on its surface by atmospheric pressure. The capsule does, in fact, collapse until the force exerted by atmospheric pressure is counter-balanced by the spring tension of the capsule. An increase of atmospheric pressure means a further collapse of the capsule, while a reduction of pressure allows the spring tension to assert itself so that the capsule opens out accordingly. The capsule movement is therefore a measure of the changing atmospheric pressure and it functions in a similar manner to the mercury column. Thus,it can be used for measuring height.

ALTIMETER

It has just been shown that an evacuated capsule forms the sensitive element of an altimeter and that the instrument is really a simple gauge for measuring atmospheric pressure. Because the total movement of the capsule is quite small over a wide range of altitudes, it is necessary to provide mechanical means of magnifying this movement. The rate of magnification required depends on the use to which the instrument is put; e.g. an altimeter used by the navigator or the flight engineer will have sufficient magnification for the instrument to be read to within, say 200 ft., while an instrument for use by the pilot would need to be read to within 10 ft., so requiring greater magnification of the capsule movement.

There are two classes of altimeter; the sensitive, and the non-sensitive types. The *non-sensitive* or simple altimeter, intended for use by the navigator, is calibrated according to the I.C.A.N. law and consists of an evacuated capsule operating a magnifying mechanism which enables the instrument to read to 10,000 ft. with a one-turn pointer movement, or 40,000 ft. with a four-turn pointer movement. In both instances there is only one pointer, which reads over a scale calibrated in thousands of feet, but in the 40,000-ft. altimeter there is also a subsidiary scale, behind the main dial, which indicates

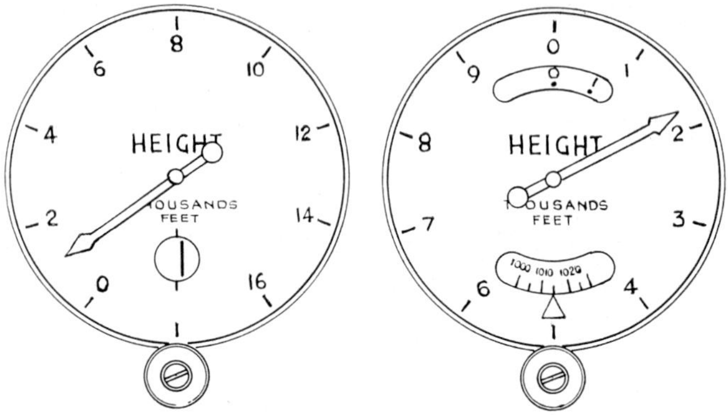

FIG. 5.3.

each 10,000 ft. mark, the pointer making one revolution for 10,000 ft. (Fig. 5.3).

In both types provision is made for checking the accuracy of the instrument in the aircraft by means of a knob which rotates the pointer, and a datum mark (or millibar scale in the case of the 40,000-ft. instrument). On setting the datum mark, or 1,013 in the case of the millibar scale, the instrument pointer will indicate a height proportional to the difference between the normal and actual atmospheric pressures.

In the 16,000-ft. altimeter rotation of the knob such as to cause the datum mark to coincide with the marks on either side of the aperture in the dial should result in the instrument reading zero only at a pressure of 1013·2 mb. When the pointer of the 40,000-ft. altimeter is set to zero by means of the setting knob, the mb scale should show the actual pressure of the atmosphere at that moment. This can be verified by taking a barometer reading and comparing it with the millibar scale reading (29·99 in. Hg = 1,013·2 mb). The mechanism of these instruments is shown in Fig. 5.4. It will be seen that the lever-length adjustment consists of moving the arm in the rocking shaft, while the lever-angle adjustment is made by movement of the capsule in the instrument frame. The knob at the front of the instru-

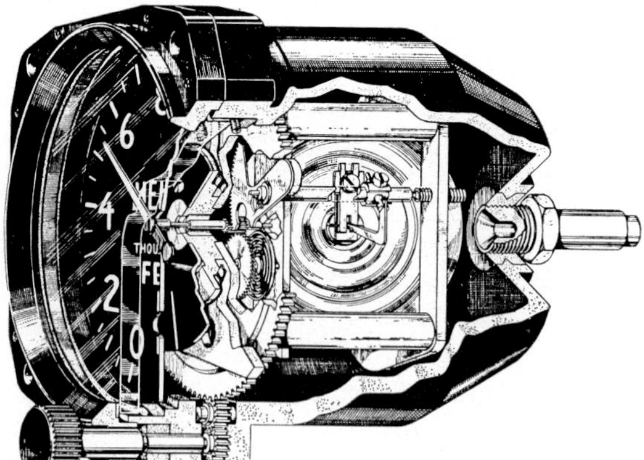

FIG. 5.4. *Non-sensitive type of altimeter.*

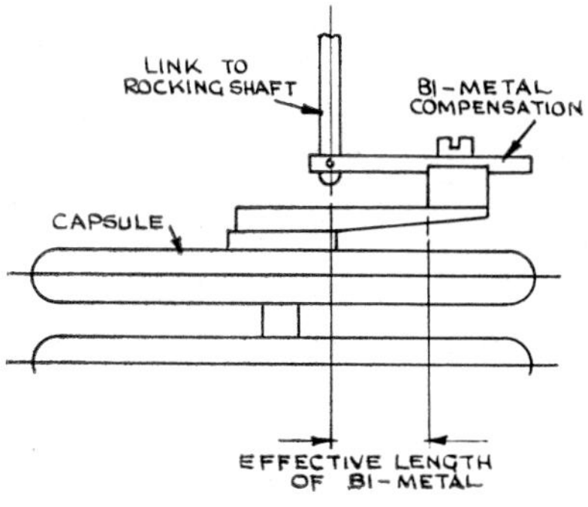

FIG. 5.5.

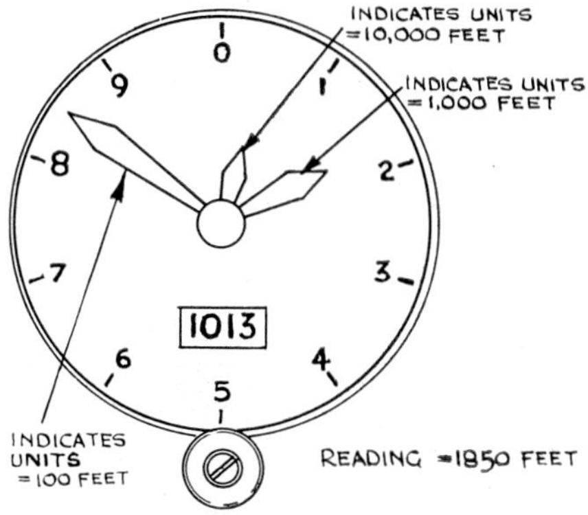

FIG. 5.6.

ment controls a pinion which meshes with gear teeth cut into the front plate of the instrument, so that rotation of the knob causes rotation of the complete mechanism and pointer. Stops are fitted to limit the movement in either direction.

Calibration. For calibration purposes the movement is placed inside a pressure chamber to which is connected a mercury barometer. A vacuum pump evacuates the pressure chamber and readings of instrument and barometer are noted at intervals throughout the range. Adjustments are then made according to the readings obtained; for example, if the instrument shows progressively increasing readings throughout the range, it is necessary to increase the lever length; i.e. to move the arm *out* of the rocking shaft. If the readings at the lower part of the scale are reasonably correct, but those at the higher end show progressively increasing errors, it is necessary to adjust the lever angle; i.e. if the readings at the top end of the scale are negative the lever angle is reduced by moving the capsule assembly *away* from the frame in which it is mounted.

If, during overhaul, the temperature-compensating device has been disturbed, it is necessary to reset it. This is done by placing the instrument in a temperature chamber and noting whether or not a change of pointer position occurs with a change of temperature. If the instrument shows a negative error outside the permitted tolerance with a reduction of temperature, it is necessary to reduce the effective length of the compensating device by moving it away from the centre of the capsule (Fig. 5.5).

Sensitive Altimeter. This altimeter is one which indicates small changes of altitude with reasonable deflections of the pointer. The principle is identical with that of the non-sensitive type in that a pressure-sensitive element detects a change in barometric pressure and operates a pointer via a gear train. Calibration of this instrument is also to the I.C.A.N. law. The gear train in the sensitive instrument must necessarily have a high ratio which requires more power to operate it. It is usual therefore to find two or more capsules in the sensitive altimeter to provide this additional power.

The gear-train movement is absorbed by no less than three pointers to provide a complete picture of the indicated altitude at any instant. A large pointer moves over the scale to indicate hundreds of feet and completes one revolution of the dial for 1,000 ft. change. A small second pointer indicates thousands of feet and completes one revolution for 10,000 ft., while a third and even smaller pointer indicates in terms of 10,000-ft. units (Fig. 5.6).

Millibar Scale. An additional feature of the sensitive type of altimeter is that it carries a subsidiary or millibar scale, the purpose of which can be explained by a simple illustration. Suppose an aircraft to be on a runway at sea level, when the altimeter should show zero height. Remember however, that the altimeter is a barometer and a change of weather conditions will be recorded by the instrument pointer, which will accordingly show a change of height although the aircraft will not have left the ground. Now consider an aircraft taking off from this same runway, with its altimeter indicating zero. As the aircraft climbs away it will be expected that the altimter will indicate the altitude at any instant. On returning to base one might expect the instrument to indicate aero on touch-down and therefore to provide some indication of the nearness of the runway in bad visibility.

It has been seen, however, that a change in conditions will cause a change in the instrument reading and therefore should such a change occur during the absence of the aircraft from base then the instrument will show this change on touch-down; for example, suppose the change of conditions reduced the pressure from its normal of 30 in. Hg to say 29·5 in. Hg. This reduction of pressure will cause an increase in the instrument reading. On page 75 it was shown that a reduction of pressure of 2 in. Hg was approximately equal to 2,000 ft.; therefore a pressure change of 0·5 in. corresponds to 500 ft. and that would be the instrument reading at the point of touch-down. If the instrument is to have any worth as a landing aid it must be supplied with a means of setting the pointers during flight to such a value that it reads zero on touch-down. The millibar scale provides this means of resetting the instrument as desired. The scale consists of a subsidiary dial, or, in some instruments, a Veeder counter which represents millibars (30 in. Hg = 1,013 millibars). The normal setting of the counter or dial is one of 1,013 at which setting—*under normal conditions*—the instrument will read zero. Should the aircraft take off from base, fly around, then return to base, the instrument must be reset before the final approach. This is done by obtaining radio information of the barometric pressure existing at the airfield on which the aircraft is to land and setting this value, by means of a knob, on the millibar counter. The instrument will then read zero on touch-down.

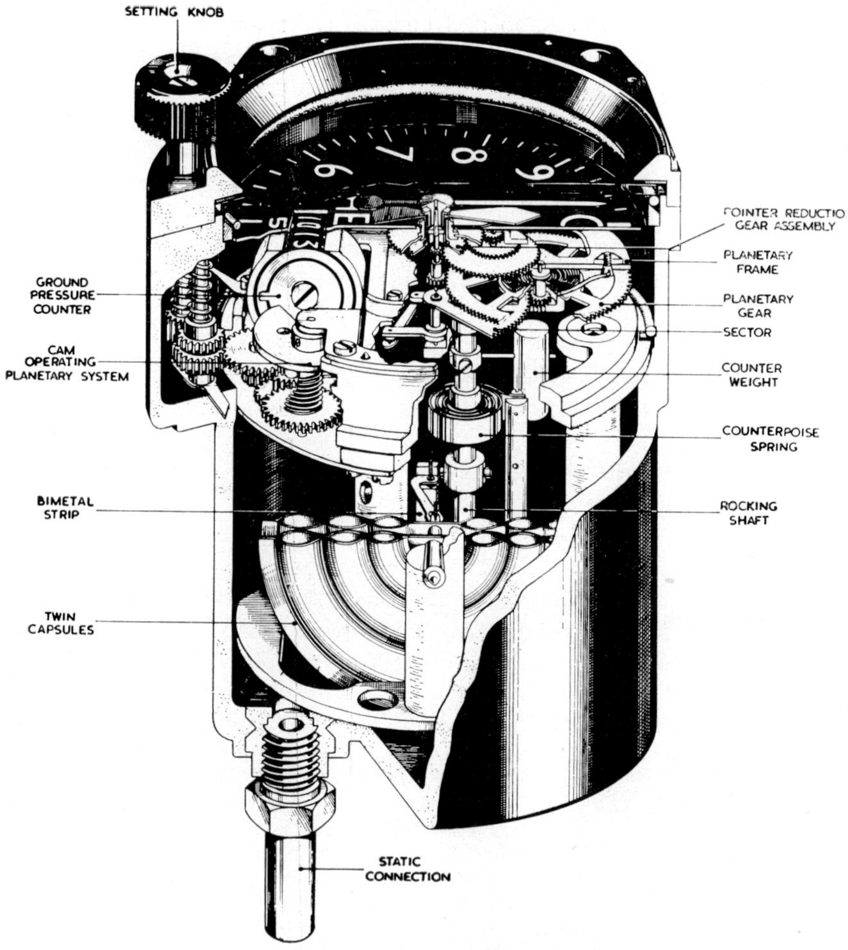

FIG. 5.7. *Sensitive type of altimeter.*

In practice there are two schools of thought on the use of the barometric scale on the sensitive altimeter, one of which follows the method set out above, and which can be summarized as follows: "Set the millibar scale to the barometric pressure at the airfield, and the instrument reads zero on touch-down." The other school of thought says that the barometric scale should be "set to a value which will ensure that the instrument reads the height of the airfield above sea level at touch-down". The first is known as a QFE setting and the second as a QNH setting; i.e. the international code letters by which the settings are recognized. In either case, adjustment of the knob means movement of the millibar scale or pointer, and the instrument pointer. Now the relationship between knob and counter is linear but the relationship between barometric pressure and altitude is non-linear and one must reconcile these two things in the sensitive altimeter.

In the "Smith" type of instrument this is done by means of a profiled cam which converts the linear movement of the counter device to a non-linear movement of the pointers. During overhaul of the instrument (Fig. 5.7) it is essential to see that the cam is inserted at the correct millibar/pointer relationship. (Details may be obtained from the manufacturer.) As an example of the non-linearity of the pressure/height curve one may observe the following approximate values:

at zero altitude 1 millibar = 30 ft.
at 10,000 ft. altitude 1 millibar = 38 ft.
at 20,000 ft. altitude 1 millibar = 51 ft.
at 30,000 ft. altitude 1 millibar = 75 ft.

Maintenance and Testing. An altimeter usually receives little maintenance beyond the periodic check by calibration against a barometer. Routine maintenance work is to see that all moving parts are free, but with minimum shake, and that all gears are clean and free from burrs and wear. The freedom of all parts from foreign matter is vital during assembly of any instrument and the altimeter is one instrument which will show up lack of cleanliness because of the large magnification ratio used in it; for example, at 60,000 ft. a movement of 0·001 in. of the capsule is enough to move the large pointer through approximately 1,000 ft. in an instrument which has been assembled with care.

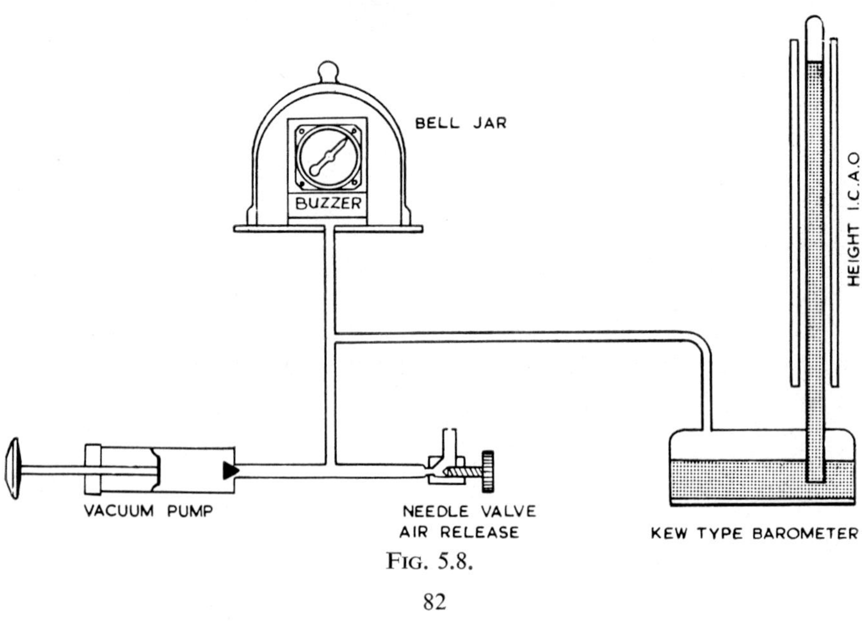

FIG. 5.8.

Table 5.1

PRESSURE/HEIGHT EQUIVALENTS

Height (*ft.*)	*Barometric Pressure* (*London Laboratory Conditions*) (*in. Hg*)
0	29·99
1,000	28·93
5,000	24·96
10,000	20·63
15,000	16·93
20,000	13·78
30,000	8·91
40,000	5·55
50,000	3·43
60,000	2·12
70,000	1·31

The altimeter is tested by comparing it with a barometer. The rig is shown in Fig. 5.8. The instrument is placed in a vacuum chamber which is connected to a barometer; on evacuating the chamber it will cause an increase in the altimeter reading and a decrease in the barometer reading. The readings are then compared and any instrument inaccuracy will be revealed. It is wise to refer to the manufacturer's literature before testing a particular instrument, but pressure/height equivalents will be as shown in Table 5.1.

Barometric Conditions. When reading the barometer, allowance must be made for temperature, latitude and height above sea level at which the test is being made (see page 75). Sometimes the corrections due to these differing factors are reduced to one correction only, known as the fiducial temperature correction. The principle of this is easy to understand; for example, suppose the correction due to temperature amounts to +15 ft., the correction due to latitude to −20 ft., and the correction due to height above sea level of the laboratory in which the instrument is being calibrated to +10 ft. The total correction is the sum of all the corrections $+ 15 + (-20) + 10 = + 5$ ft. Now by controlling temperature, this could be reduced to zero. However, that would be difficult in practice and therefore use is made of known values, as indicated in Tables 5.2 and 5.3.

An example will serve to illustrate the use of these tables; suppose the barometer to be housed at a latitude of 35 deg. and at a point 2,000 ft. above sea level. Looking down the left-hand column of Table 5.2 to 35 deg. then opposite this value under the column headed 2,000 ft. is a figure 7·7. This represents the fiducial temperature in degrees C., i.e. the temperature at which

Table 5.2

FIDUCIAL TEMPERATURE IN DEGREES C. FOR VARYING LATITUDE AND HEIGHT ABOVE MEAN SEA LEVEL

Latitude (Deg. N or S)	*Height of Station in feet above Mean Sea Level*							
	−500	0	1,000	2,000	3,000	4,000	5,000	10,000
90	28·2	28·0	27·7	27·3	27·0	26·7	26·3	24·6
80	27·3	27·1	26·8	26·5	26·1	25·8	25·4	23·8
70	24·8	24·6	24·3	23·9	23·6	23·3	22·9	21·2
65	23·0	22·8	22·5	22·1	21·8	21·5	21·1	19·4
60	20·9	20·7	20·4	20·0	19·7	19·4	19·0	17·3
55	18·6	18·4	18·1	17·7	17·4	17·1	16·7	15·0
54	18·1	17·9	17·6	17·3	16·9	16·6	16·2	14·6
53	17·6	17·4	17·1	16·8	16·4	16·1	15·8	14·1
52	17·1	16·9	16·6	16·3	15·9	15·6	15·3	13·6
51	16·6	16·4	16·1	15·8	15·4	15·1	14·8	13·1
50	16·1	15·9	15·6	15·3	14·9	14·6	14·3	12·6
45	13·6	13·4	13·1	12·7	12·4	12·1	11·7	10·0
40	11·0	10·9	10·5	10·2	9·9	9·5	9·2	7·5
35	8·6	8·4	8·1	7·7	7·4	7·1	6·7	5·0
30	6·3	6·1	5·8	5·4	5·1	4·8	4·4	2·7
25	4·2	4·0	3·7	3·3	3·0	2·7	2·3	0·7
20	2·4	2·2	1·9	1·6	1·2	0·9	0·5	−1·1
15	0·9	0·8	0·4	0·1	−0·2	−0·6	−0·9	−2·6
10	−0·1	−0·3	−0·6	−1·0	−1·3	−1·7	−2·0	−3·7
5	−0·8	−1·0	−1·3	−1·6	−2·0	−2·3	−2·7	−4·3
0	−1·0	−1·2	−1·5	−1·9	−2·2	−2·5	−2·9	−4·6

the barometer would have to be kept so as to give true readings. If the actual temperature was 16·7 deg. C. then the fiducial temperature correction will be proportional to +9 deg. C. The correction to be applied to the barometer can be obtained from Table 5.3.

Now if the check point of the instrument is 1,000 ft. this corresponds to a pressure of 28·93 in. Hg under London Laboratory conditions. Under the conditions quoted above the correction per degree is 0·0051 in., therefore the actual amount to be added to the observed reading is $+9 \times 0{\cdot}0051$ in. $=$ $+0{\cdot}0459$ in.; i.e. at 1,000 ft. the barometer should read $28{\cdot}93 + (9 \times 0{\cdot}0051)$ $= 28{\cdot}9759$ in. Hg. Any difference from this reading represents instrument error. Should the fiducial temperature value obtained from Table 5.2 be above the actual temperature, the correction would have to be subtracted from the values given in column 2 of Table 5.3.

The mechanism is adjusted by altering the lever length and lever angle. The lever length is adjusted by moving the lever in or out of the rocking shaft,

Table 5.3

BAROMETER CORRECTIONS

Height (I.C.A.N. feet)	*Inch Scale Reading under London Laboratory Conditions*	*Corrections* (*in.*) *for each* 1 *deg. C. from Fiducial Temperature*
−1,000	31·09	0·0055
0	29·99	0·0053
1,000	28·93	0·0051
2,000	27·89	0·0050
5,000	24·96	0·0044
10,000	20·63	0·0037
15,000	16·93	0·0030
20,000	13·78	0·0025
25,000	11·13	0·0020
30,000	8·91	0·0016
35,000	7·06	0·0013
40,000	5·55	0·0010
45,000	4·37	0·0008
50,000	3·43	0·0006
55,000	2·70	0·0005
60,000	2·12	0·0004
65,000	1·67	0·0003
70,000	1·31	0·0002

while the lever angle is changed by moving the capsule stack towards or away from, the rocking shaft.

RATE-OF-CLIMB INDICATOR

The rate-of-climb indicator (sometimes known as a vertical speed indicator) indicates a rate of change of height. Since the altimeter indicates height by measuring the barometric pressure, then it can be imagined that the rate-of-climb indicator in fact measures rate of change of barometric pressure. It does this by measuring the differential pressure across a leak through which a flow of air, proportional to rate of change of height, is passing.

The calibrated leak assembly is mounted in an airtight case having a low-pressure union at the rear of the case to connect the instrument to a static pressure line. The pressure in this line is the static atmospheric pressure surrounding the aircraft. Connected to one side of the leak is a capsule, while the other side of the leak is open to the interior of the case (Fig. 5.9). Considering an aircraft at a given constant barometric height, the pressure in the capsule will be that of the atmosphere around the aircraft. Similarly, the leak assembly will have allowed this same pressure to exist inside the case and therefore to act on the *outside* of the capsule, the net result being that there is

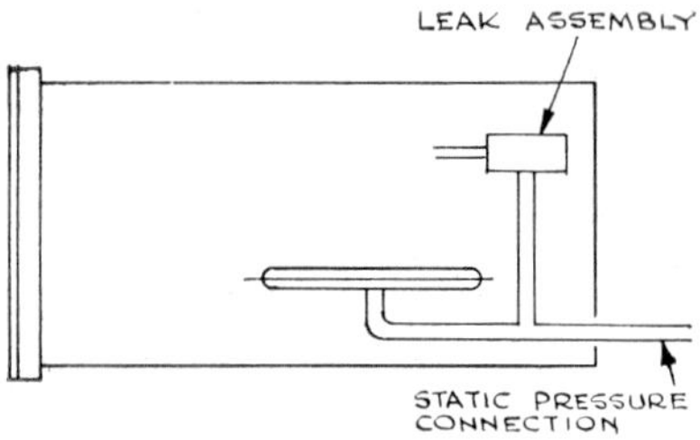

FIG. 5.9.

no difference of pressure between the inside and the outside of the capsule and no resulting movement from it.

If the aircraft altitude is reduced there is an increase in the barometric pressure which is felt immediately *inside* the capsule. At the same time, this increased pressure will slowly pass through the leak assembly to the interior of the case to act on the *outside* of the capsule. The leak assembly therefore introduces a time lag in the equalization of pressures and during this time lag the capsule will respond to the differential pressure and actuate a pointer.

It will be appreciated that if the aircraft maintains a constant rate of descent there is a constant rate of increase of pressure in the capsule and, because of the leak assembly, a constant rate of increase of pressure in the case. However, the time factor of the leak causes the pressure in the case to lag behind the pressure in the capsule and hence the differential pressure is maintained and the instrument pointer will indicate a constant rate of descent. As soon as the aircraft levels off, the pressure in the capsule remains constant and the pressure in the case becomes equal to the atmospheric pressure in the capsule, thus returning the pointer to zero.

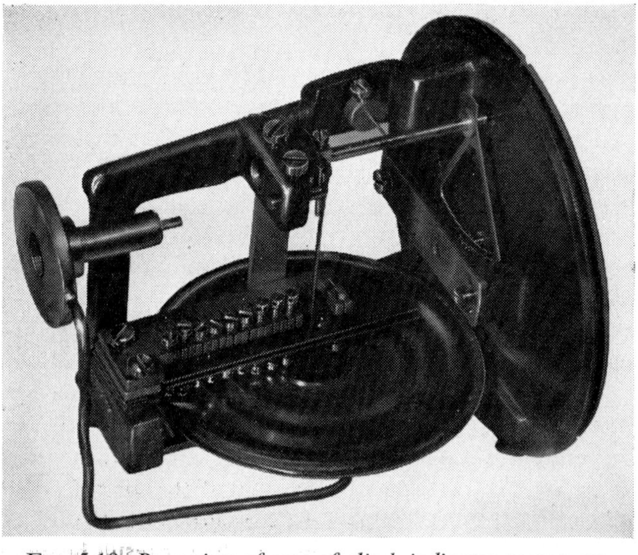

FIG. 5.10. *Rear view of rate-of-climb indicator movement.*

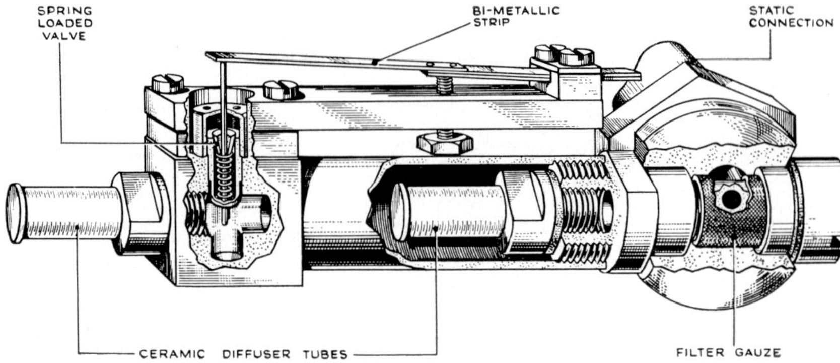

FIG. 5.11.

The instrument consists of two main parts: a pressure gauge, and a calibrated leak. The pressure gauge is of the simple capsule type and can be calibrated as such if the leak is sealed off. The differential pressures measured are quite small, being in the region of 20 mm H_2O for full-scale deflection. The calibration mechanism of the instrument is similar to that of the sensitive airspeed indicator, comprising a tuning spring which is adjusted by screws to vary the spring rate (Fig. 5.10). The type of calibration leak varies according to make but there are two main types: the ceramic leak, or porous-pot type, and the orifice and capillary type. The ceramic leak is simply a porous ceramic tube which is mounted in the case so as to be between the static line and the case interior. The flow of air to and from the case is controlled by the degree of porosity of the tube. The rate of flow of air, however, also depends on altitude and viscosity of the air passing through the tube. It is necessary therefore to provide some means of compensation for changes of altitude and viscosity. The leak assembly with compensating device is shown in Fig. 5.11; it contains two similar ceramic tubes in series, with a valve between them.

The action of the device is as follows. Under conditions of lower altitude and normal viscosity the flow of air through the first tube is allowed to pass through the valve to the interior of the case. At increasing altitudes and lower viscosity the pressure drop across the leak will be reduced, but a bi-metal arm on the assembly is affected by the reduction of temperature with increasing altitude, and operation of the arm closes the valve. The result is that air must now pass through both ceramic tubes to the case interior and so the pressure drop across the assembly is maintained, thus providing compensation for the variation in altitude and viscosity. In practice, the valve assumes intermediate positions between open and closed, but the net result of the various positions is a leak value which gives a constant differential pressure for a fixed rate of climb or descent at any altitude.

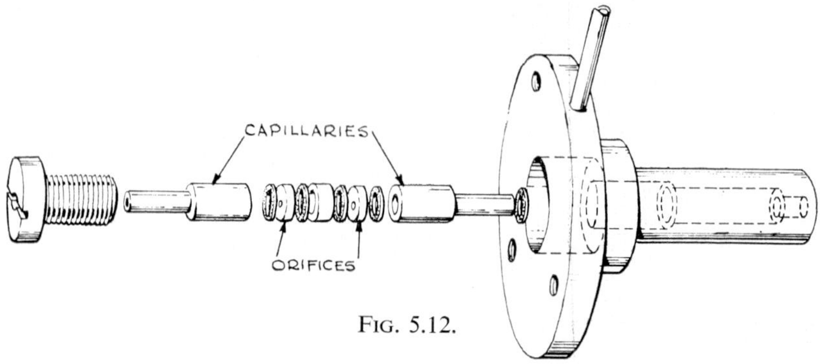

FIG. 5.12.

Orifice and Capillary Type. This comprises two separate leak assemblies, one an orifice plate and one a capillary tube, but mounted together in series with no moving parts (Fig. 5.12). The principle employed here is that both an orifice and a capillary tube would act as a leak as required by the instrument. If either were used alone, however, it would be necessary to have some device, such as is used with ceramic tubes, to achieve compensation. In practice, because the flow through a capillary tube is regarded as laminar and that through an orifice as turbulent, the two devices have opposite characteristics. Therefore, using both together in series, the effect of one offsets the effect of the other and again a constant pressure differential is obtained for a given rate of climb or descent, at any altitude.

The instrument is tested by using a stop-watch and a barometer, the procedure being as follows. The instrument is placed in a pressure chamber which is connected to a mercury barometer. The pressure in the chamber is reduced until the instrument reads, say 2,000 ft./min. climb, and the fall of mercury in the barometer is timed through a height of 2,000 ft. Obviously, if the indicator is correct the time of fall of the mercury should be exactly one minute. Any deviation from this period represents instrument error; e.g.:

(1) Instrument set to read 2,000 ft./min.

(2) Time for mercury level in barometer to change 2,000 ft. = 58 seconds.

(3) Actual rate of change of height = 2,000/58 × 60 = 2,070 ft./min.

(4) Instrument therefore under-reads by 70 ft./min.

It is usual to check the rate-of-climb indicator using the beginning of the barometer scale (i.e. lower altitude) since at the end of the barometer scale (higher altitude) the effects of high altitude would obtain without the appropriate change of temperature for compensation purposes.

As with other instruments connected to the pitot/static system it is usual to check that the case is airtight. This is done in a similar manner to that described for airspeed indicators on page 61.

CHAPTER 6

Measurement of Temperature

THE measurement of temperature on aircraft is mainly concerned with engine operation but one exception to this is that of air-temperature measurement. The various thermometers used in temperature measurement on a modern aircraft are shown in Table 6.1.

The *oil-temperature thermometer* gives an indication of the lubricating-oil temperature in the aircraft engine so that optimum working conditions for the engine can be maintained. The lubricating oil is pumped into the engine under pressure from a reservoir and returned to the reservoir via a cooling radiator. The oil-temperature thermometer is usually inserted in the oil system at a point close to the engine. The *coolant thermometer* is inserted in the cooling system at the "header" tank from which the liquid coolant passes to the engine for cooling purposes. Coolant returns from the engine via a radiator with variable shutters to control the degree of cooling. Thus the coolant

Table 6.1

THERMOMETERS

Purpose	*Type*	*Range (deg. C.)*
Oil thermometer	Mechanical or electrical	0 – 100
Coolant thermometer	Mechanical or electrical	40 – 140
Cylinder thermometer	Electrical	0 – 350
Exhaust-gas thermometer	Electrical	0 – 800
Air-temperature thermometer	Mechanical or electrical	−70 to +50

thermometer gives an indication of the position in which to set the radiator shutters, which are remotely controlled from the aircraft cabin.

The *cylinder thermometer* is used on air-cooled engines to give an idea of the operating conditions and is, therefore, the equivalent of the coolant thermometer used for liquid-cooled engines. The cylinder thermometer gives an indication of the setting required for the cowl-gills which control the cooling of the air-cooled engine. In jet aircraft, operating conditions are governed by the temperature of the exhaust gases from the turbine. The *exhaust-gas thermometer*, or thermometers will be found in the tail-pipe. Air-temperature measurements are concerned with cabin air, in the case of air-conditioned aircraft, and temperature of the outside atmosphere. In the latter case the indication of the instrument is used when applying corrections to indicated airspeed and altitude for use in navigational purposes.

Temperature and Heat. Before discussing the detailed mechanisms of the instruments which are used to measure temperature, it should be made clear that *temperature* and *heat* are not different ways of expressing the same thing. *Temperature* expresses the *degree of hotness* or *coldness* of a body; i.e. if a thing feels cold to the touch then it can be said to be at a lower temperature than the hand which is touching it. *Heat* is, however, a form of energy which is not related to temperature; for example, an egg-cupful of water when brought to boiling point has a high temperature, but, because of the small volume of the egg-cup, very little energy is required to raise the water to this high temperature. The *energy* contained by this amount of hot water is therefore very small. Now, if the same amount of energy were applied to a bathful of water, then it would be found that the water would be cold to the touch; i.e. in each case, the amount of energy applied is the same but the temperatures are very different. In the measurement of hot and cold bodies it is the temperature with which we are concerned and not the heat energy.

MERCURY THERMOMETER

The most familiar form of temperature-measuring instrument is the simple mercury-in-glass thermometer. This consists of a glass tube of very fine bore, at one end of which is a thin-walled reservoir containing mercury. The tube is filled under vacuum or at a high temperature to exclude air. Mercury is sensitive to changes of temperature, so that a change in level in the tube will indicate a temperature change.

To be able to compare temperatures it is necessary to have a scale marked on the thermometer. The two common scales are Fahrenheit and Celsius; both scales use common fixed points, but the division of the scales varies one from the other. One fixed point is the "ice point"; e.g. the point at which water and ice exist together. Another fixed point is the "steam point" or the temperature of steam from pure distilled water. The temperature difference between these two fixed points is known as the *fundamental interval.*

The Fahrenheit scale is one in which the fundamental interval is divided into 180 equal parts or degrees. The lower point of this interval is designated 32 deg. F. and the upper point 212 deg. F. The zero of the Fahrenheit scale is accordingly 32 Fahrenheit degrees below "ice-point". The Celsius scale is one in which the fundamental interval is divided into 100 equal parts or degrees, the lower point being designated 0 deg. C. and the upper point 100 deg. C. A comparison of scales indicates that 100 Celsius degrees are equivalent to 180 Fahrenheit degrees, showing that 5 Celsius degrees are equivalent to 9 Fahrenheit degrees. Bearing this relationship in mind, it is possible to convert a temperature reading in Fahrenheit degrees to Celsius degrees; e.g. if a thermometer indicates a temperature of 59 deg. F. the equivalent temperature in Celsius degrees may be determined as follows. Since the Fahrenheit scale indicates a temperature of 32 deg. at the ice-point, then 59 deg. F. means 27 deg. F. above the ice-point. It has just been shown that 9 Fahrenheit degrees equal 5 Celsius degrees, therefore the Celsius scale will indicate $27/9 \times 5 = 15$ deg. above ice-point. But the ice-point on the Celsius scale is zero, therefore a temperature of 59 deg. F. is the same as 15 deg. C. The conversion is usually shown as follows:

$$\text{deg. F. to deg. C.} = 5/9\,(\text{deg. F.} - 32)$$
$$\text{deg. C. to deg. F.} = 9/5\,(\text{deg. C.}) + 32$$

Calibration. The mercury-in-glass thermometer is calibrated by immersing it in a mixture of ice and water to obtain the lower fixed point and thereafter in steam from distilled water to obtain the upper fixed point. The scale is then uniformly divided according to whether the Fahrenheit or Celsius scale is required. The thermometer should always be immersed to the same point to obtain accurate readings and this point is usually marked on its scale.

To calibrate the thermometer outside the range of 0–100 deg. C. (i.e. the fundamental interval) a number of other fixed points are laid down in an internationally agreed scale. These are:

Oxygen point	−182·97 deg. C.
Sulphur point	444·60 deg. C.
Silver point	960·80 deg. C.
Gold point	1,063·00 deg. C.

A number of "secondary fixed points" are also agreed ranging from: gaseous carbon dioxide −78·50 deg. C., to melting point of tungsten 3,380·00 deg. C.

Mercury normally boils at approximately 357 deg. C., so that a mercury thermometer which is to be used for higher temperatures must be filled at high pressure, thereby increasing the temperature of the boiling point. (The boiling point of any liquid is dependent on pressure, and increasing pressure increases the boiling point.)

MERCURY-IN-STEEL THERMOMETER

The mercury-in-glass thermometer is not practicable for use in aircraft although it is used as a standard against which aircraft thermometers are checked. A more practical temperature-measuring instrument is the mercury-in-steel thermometer. This comprises a steel bulb, capillary tube and Bourdon tube completely filled with mercury. A change of temperature causes a change of volume of mercury and the Bourdon tube moves to show the change. A pointer moves over a scale to indicate the temperature, usually in deg. C. in the case of aircraft thermometers (Fig. 6.1.).

Compensation for Errors. Errors in this instrument arise mainly from the fact that the mercury in the capillary tube and indicator is also sensitive to temperature changes and these will be registered on the dial. To reduce the errors to a minimum, the volume of mercury in the capillary tube and Bourdon tube is kept small in comparison with the volume of the bulb. Compensation for remaining errors is effected in several ways, a common method being as follows. Compensation for temperature errors in the capillary tube can be achieved by running a fine Invar wire down the length of the capillary so that the cross-section of the tube would be as in Fig. 6.2. The effect of this is that an increase of temperature of the mercury in the capillary tube causes an increase of volume which would normally show as an increased reading on the indicator. However, because of the differential expansion of the steel tube and Invar wire (the expansion of which remains practically constant), an increase in the volume of the space between the wire and the walls of the tube occurs and this accommodates the increase in volume of mercury. Compensation in the indicator can be arranged by fitting a bi-metal strip to the mechanism, which counteracts the effect of changes of volume of mercury in the indicator.

Double-Spiral Bourdon Tube. In the double-spiral Bourdon-tube instrument there is no mechanism, and compensation in this case takes the form of a bi-metal helix which forms the last coil of the double-spiral tube (Fig. 6.3).

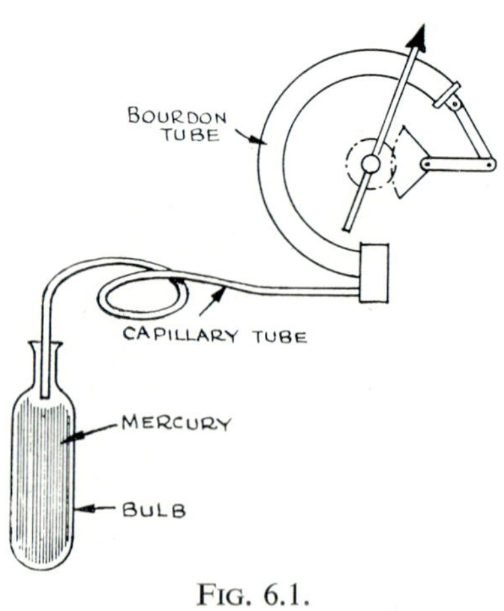

FIG. 6.1.

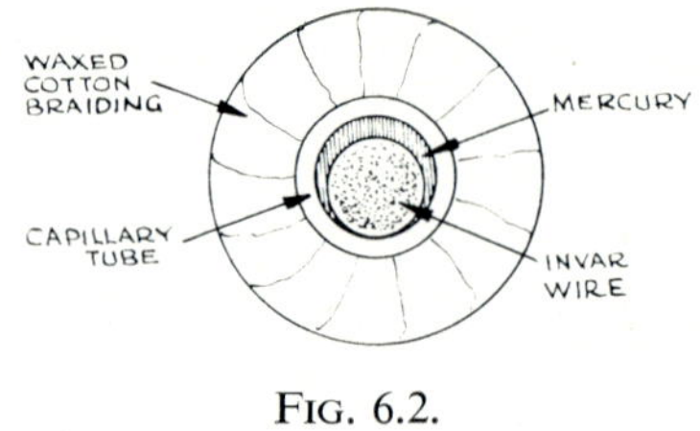

FIG. 6.2.

There is little maintenance that can be done on this instrument, and because of its construction as a system comprising bulb, capillary tube and Bourdon tube, it is difficult to remove from the aircraft. Calibration checks, can, however, be carried out by immersing the bulb in an oil bath together with a standard mercury-in-glass thermometer. Readings are compared as the oil bath is slowly heated, care being taken to agitate the bath so as to obtain a uniform temperature before taking any readings. The errors of the instrument are in the region of 2 per cent.

The main use of the mercury-in-steel thermometer is as an oil-temperature gauge indicating the temperature of the engine lubricating oil. It is also used as an air-temperature gauge for indicating the temperature of air outside the

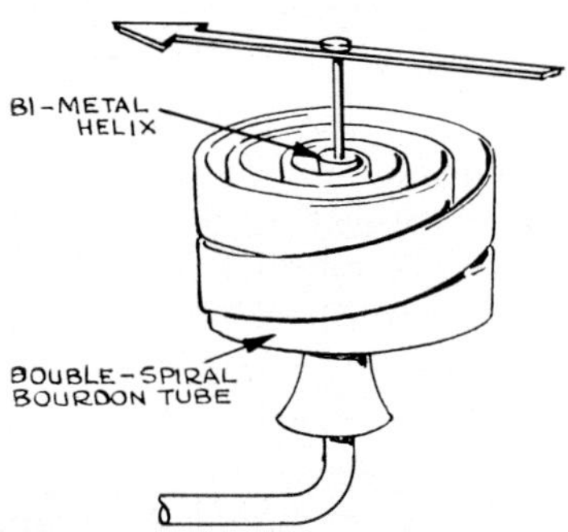

FIG. 6.3.

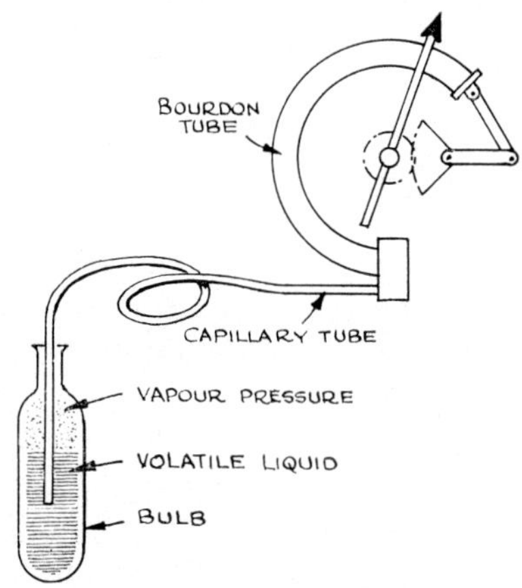

FIG. 6.4.

aircraft, so that the navigator has a means of applying corrections to air-speed, altitude and so on. On larger modern aircraft these mechanical temperature gauges have been superseded by the electrical types of thermometer (see Chapter 11) which have the advantage of easier maintenance because the various units can be disconnected from a terminal block.

VAPOUR-PRESSURE THERMOMETER

Another mechanical temperature gauge, or thermometer, is the vapour-pressure thermometer. This comprises a thermometer bulb, capillary tube and Bourdon tube as in the mercury system, but the operating medium is a highly volatile liquid such as ethyl ether. The volume of liquid contained in the system is not sufficient completely to fill it and the remaining space contains vapour given off by the chosen liquid (Fig. 6.4).

A rise in the temperature of the thermometer bulb increases the evaporation rate of the liquid to produce a greater vapour pressure. This pressure is transmitted, via the liquid, to the Bourdon tube, thus indicating the value of the pressure. Since the pressure is proportional to the bulb temperature, the indicator can be calibrated in terms of Fahrenheit or Celsius degrees. The scale range of this instrument is limited by the fact that the vapour pressure is negligible at normal temperatures, and a common range for this instrument is +40 to +140 deg. C. In this form the instrument is often used as a coolant thermometer, the bulb being immersed in the radiator header tank.

The instrument is calibrated in the same manner as the mercury-in-steel thermometer; i.e. the bulb is immersed in a tank containing a master mercury-in-glass thermometer. The tank is heated and the readings of the vapour-pressure gauge are compared with the readings of the mercury thermometer. Maintenance is the same as for normal pressure gauges, although care should be taken in handling the system, particularly in respect of the capillary tube. The slightest "crack" in the system, or a system where the capillary tube is not properly soldered into the bulb and indicator, will allow evaporation of the fluid. This results in a larger space being left to be filled by the vapour and consequently a lower pressure indication for a given temperature. This is not an unusual fault with this type of instrument.

In the mercury type of thermometer it was shown that the system comprising bulb, capillary tube and Bourdon tube was made of steel. This is because mercury forms an amalgam with most other metals. There is no necessity for such a system in the case of the vapour-pressure thermometer and the bulb is usually of brass, while the capillary is of annealed copper; the Bourdon tube may be phosphor bronze or beryllium copper.

The principles and operation of electrical thermometers are described in Chapter 11.

CHAPTER 7

Fuel and Oil Contents Gauges

THE measurement of the quantity of fuel or oil by mechanical means is not particularly precise but the methods used for this form of measurement are simple. Basically, most of these methods consist of a device which indicates the level of liquid in a tank, since this determines the quantity. The level indicator is usually a cork float well coated with varnish to prevent saturation. The float rides on the surface of the liquid, so changing its position in the tank as the quantity of liquid varies. Movement of the float is conveyed to a scale which indicates the quantity of liquid in the tank.

Typical Movements. In one type, a float is connected to an arm which is pivoted so as to allow the float to move. The movement of the float arm is translated into the rotational movement of a spindle by means of a gear drive.

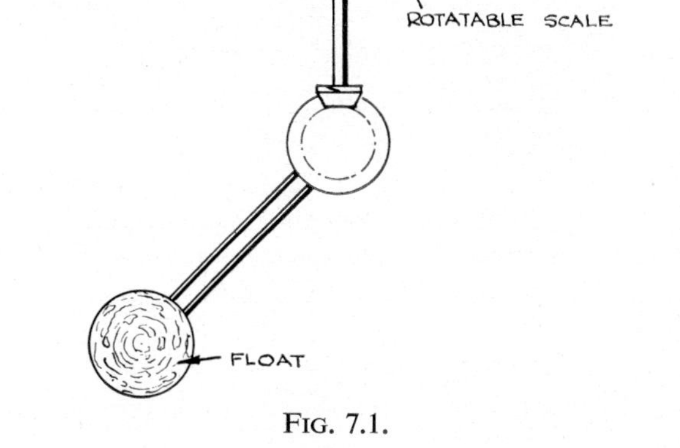

FIG. 7.1.

At the top of the tank is mounted an inverted metal cup having a scale calibrated in gallons (Fig. 7.1). A transparent cap, forming a seal to the tank, fits over the scale.

A more effective method of sealing the tank is to have an arrangement similar to that described in the preceding paragraph, but in which the rotating spindle (Fig. 7.2) carries a permanent magnet; above the magnet there is a metal seal. The lower magnet, driven by the spindle, acts on the upper magnet by virtue of the magnetic flux linking them, so rotating the upper magnet in a like manner, and moving the pointer accordingly.

Where the tank will not allow sufficient movement of a long float arm, a spiral device is used, as shown in Fig. 7.3. In this case the float is contained in

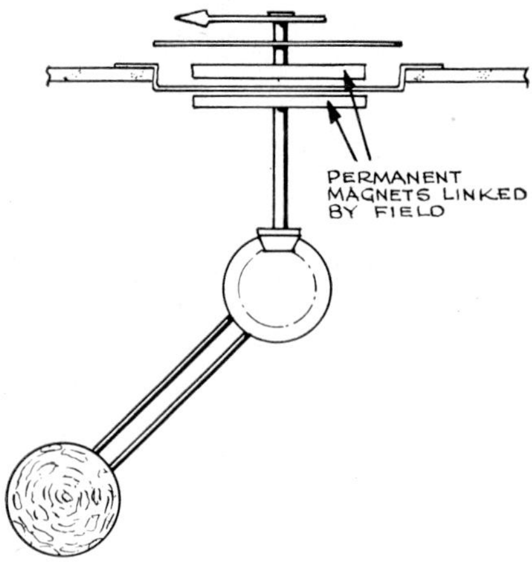

FIG. 7.2.

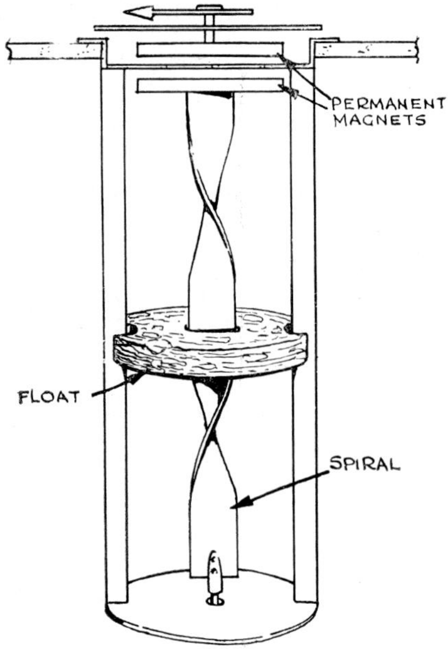

FIG. 7.3.

a long tube on the inside of which are two metal runners arranged to contain the float as it moves up and down when the level of the liquid changes. In the centre of the float is a slot through which passes a spiral strip, the action being so contrived that with varying positions of the float the spiral moves to some angular position which is a measure of the float level. This angular movement is transmitted to a pointer by means of a magnetic link.

Errors. The errors associated with the foregoing instruments are: position error; friction error; error caused by wrong installation; loss of magnetism. Position error is the main source of error and is due to the position

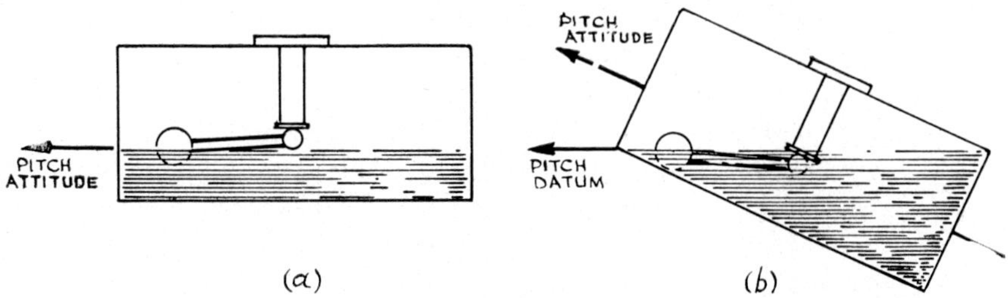

FIG. 7.4.

of the tank at the time of reading. It is, therefore, dependent on aircraft pitch attitude. Consider a tank such as the one shown in Fig. 7.4(*a*) which is half full, with the aircraft flying straight and level. It will be seen that the float arm is at an angle less than 90 deg. with the operating spindle. Now consider the same tank arrangement with the aircraft in a nose-up attitude (Fig. 7.4*b*). The tank is still half full but the float is now making an angle greater than 90 deg. with the operating spindle and this makes an appreciable error in reading the scale. It will be seen that the state of trim of the aircraft will determine the magnitude of the position error. Any friction in the gear drive between float arm and spindle will result in the float riding high or low, respectively, as the tank is emptying or filling. The indicator is made to fit, and is calibrated for, a prototype tank. Any deviation in tank structure from the prototype must result in an error proportional to the manufacturing tolerance. In the case of instruments with a magnetic link, any friction in the pointer spindle will result in a small error which could vary if the magnet strength varies with age.

Calibration. The most accurate method of checking the calibration of a fuel gauge of this type is first to drain the tank. With the aircraft in the normal flying attitude, measured quantities of fuel are poured into the tank and each indicator reading is observed. Any difference between measured and indicated values shows the instrument error. No calibration adjustments are available on this type of mechanism, but stops are provided to limit the position of the float arm between minimum and maximum readings. It is usually possible to adjust these stops if required.

HYDROSTATIC FUEL GAUGE

The hydrostatic fuel gauge is one instrument in which there are no moving parts in the fuel-tank attachment. The instrument is similar to the transmitting type of pressure gauge described in Chapter 3 and comprises a transmitting unit, capillary tube and Bourdon-tube gauge. The transmitting unit is connected to the bottom of the tank so that fuel enters the transmitter. The "head" of fuel above the transmitter creates a pressure which is conveyed via the capillary tube to the Bourdon-tube gauge. This is of conventional design, but effectively is a differential pressure gauge; that is, the pressure inside the Bourdon tube comprises the pressure due to the head of liquid plus the atmospheric pressure above it. Therefore it is necessary, in order to read only the pressure of the liquid, to convey the pressure of the atmosphere acting on the fluid to the interior of the case in which the Bourdon tube operates. In this way the reading on the dial is that due to the pressure of the liquid in the tank (Fig. 7.5).

This instrument also suffers from position error in that changes of aircraft trim will affect the head of fuel above the transmitter and so vary the reading on the gauge. Additionally, since the gauge is reading *pressure*, which is proportional to hd (where h is head of liquid and d is the density of liquid),

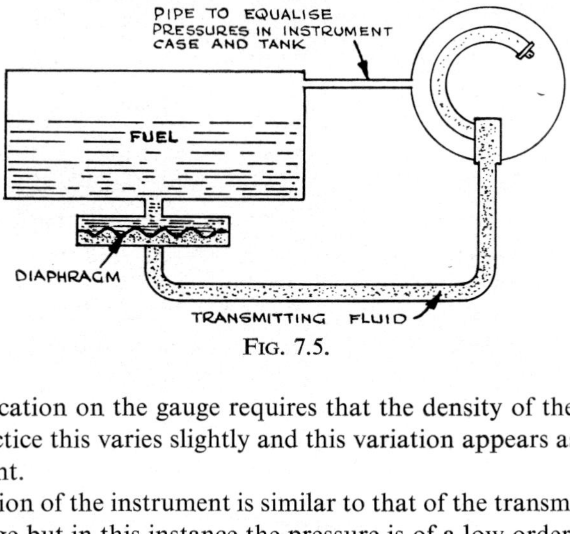

FIG. 7.5.

then the indication on the gauge requires that the density of the fuel be constant. In practice this varies slightly and this variation appears as an error on the instrument.

Calibration of the instrument is similar to that of the transmitting type of pressure gauge but in this instance the pressure is of a low order. This means that a water or mercury manometer would be used as the "standard" against which the instrument is to be calibrated.

FUEL SWITCHES

In addition to the fuel gauges already mentioned, fuel switches may also be employed in modern aircraft. The simplest of these is the fuel-level switch. This has a pivoted cork float carrying a permanent magnet (Fig. 7.6). In a certain position of the float the magnet attracts a soft-iron armature mounted on a spring contact. The spring contact can be used to make or break an electrical circuit so operating a lamp indicator; for example, if the switch is mounted in the vertical position, the float hangs downwards and the soft-iron armature is held by a secondary magnet.

(*Right*) FIG. 7.6. *Fuel level switch with pivoted cork float.*

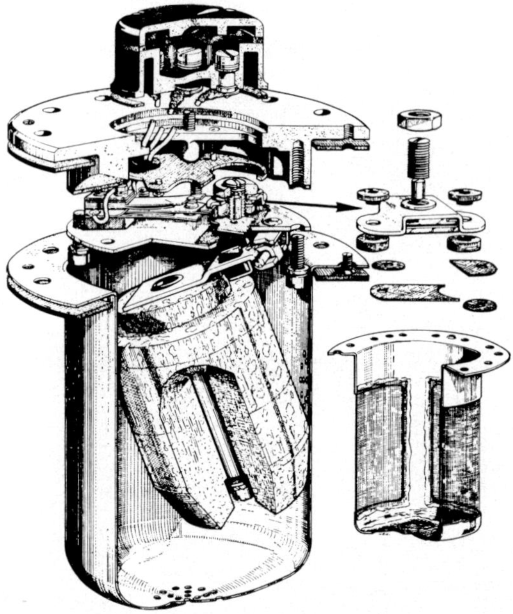

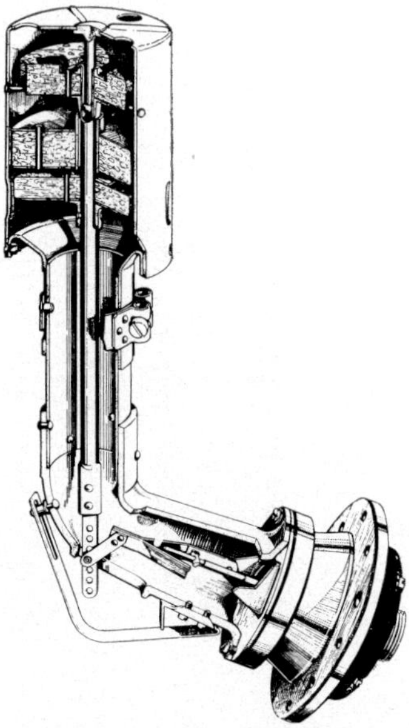

(*Left*) FIG. 7.7. *Fuel cut-off switch.*

With rising fuel level, as in refuelling, the cork float will rise and the magnet associated with the float will overcome the pull of the secondary magnet. The iron armature moves and a circuit is completed to light a warning indicator lamp to show that refuelling of the particular tank is complete. Similarly, if the assembly is mounted in an inverted position, at the bottom of the tank, the float will be forced upwards and the electrical contact will be open. As the fuel level falls the float will fall accordingly, the contact will be closed by the attraction of the iron armature, and a warning light will be switched on to indicate a minimum tank level.

Two-stage Fuel Cut-off Switch. This is a device built into a tank for the purpose of controlling refuelling (Fig. 7.7). Although a little more complicated in construction than the fuel-level switch, the principle of operation is simple; that is, during refuelling the two-stage cut-off switch will indicate that the tank is nearly full by operating the first stage of the device. This is a switch which controls the refuelling pump and reduces the rate of fuel flow into the tank. As the tank slowly fills, the second stage of the switch operates, this time to shut off the refuelling pump as the tank is filled to its normal capacity.

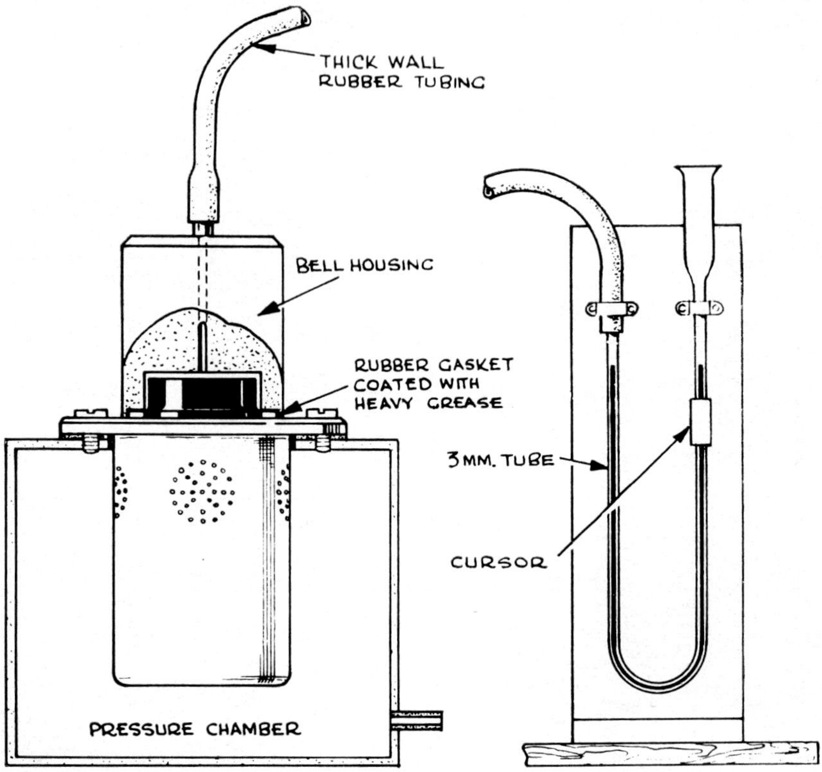

FIG. 7.8.

Leak Testing. In order to test for possible leaks in contents gauges or switches it is usual to mount the units concerned in a pressure chamber (Fig. 7.8). An air pressure of 15 lb. per sq. in. is then applied to the chamber and any leak should be noted. This will be indicated on the water manometer.

Note. The simple fuel-gauging instruments mentioned at the beginning of this chapter are only used nowadays on the smaller types of aircraft. The modern airliner, however, still makes use of the fuel switches described here, although its fuel-gauging system is of a more precise nature, as outlined in Chapter 15.

CHAPTER 8

Measurement of Rotational Speeds

MEASUREMENT of rotational speeds demands a knowledge of angular velocity, or of distance and time. There are two common mechanical instruments used on aircraft for measuring rotational speeds (i.e. engine speed) and these are: centrifugal governor-type instrument; chronometric instrument. In both instances the instrument is mounted in the aircraft cabin and is therefore remote from the engine which operates it. The connecting link between indicator and engine is a flexible drive which is driven by the engine and which operates the indicator. To increase the efficiency of and to reduce wear on the flexible drive, it may be operated at some sub-multiple of the engine speed. The indicator may contain a gear having a ratio which produces in the instrument a rotational speed equal to that of the engine which is operating it.

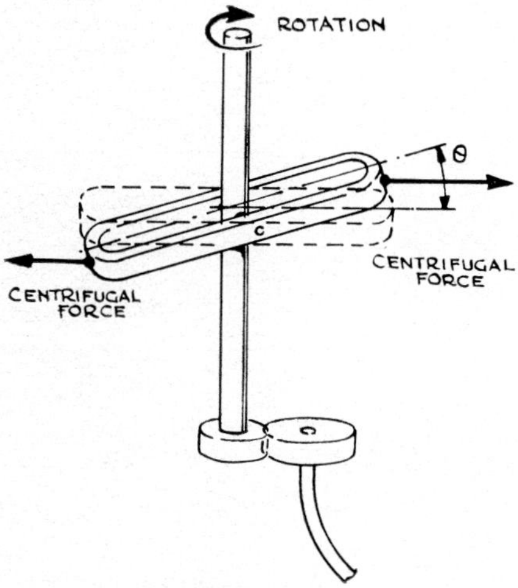

FIG. 8.1.

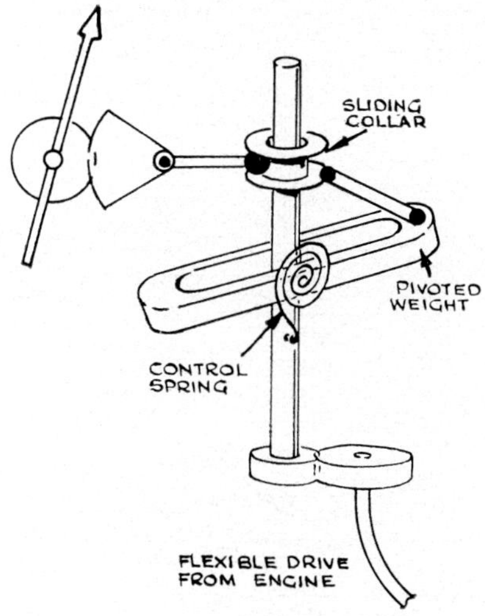

FIG. 8.2.

CENTRIFUGAL GOVERNOR-TYPE INSTRUMENT

In this instrument the flexible drive rotates a shaft (Fig. 8.1) on which is pivoted an arm. At each end of the arm is a small weight. The effect of centrifugal force, created by the rotation of the shaft, is to force the weights outwards until they are at right-angles to the shaft. If the arm is spring-restrained then a balance is reached between the value of centrifugal force for a particular rotational speed and the force exerted by the spring. The angular movement θ is therefore a measure of the rotational speed. This angular movement is transmitted via a link (Fig. 8.2) to a collar mounted on the rotating shaft. The collar in turn drives another link which operates a quadrant and pinion to drive a shaft on which the pointer is mounted. The direction of shaft rotation does not affect the operation of the indicator. Adjustments are made on this instrument by increasing or decreasing the spring tension; e.g. if the instrument over-reads it is necessary to increase the spring tension, while an under-reading is corrected by reducing the spring tension.

CHRONOMETRIC ENGINE SPEED INDICATOR

In this instrument the angular velocity of a rotating shaft is measured over a short period of time. An integrating mechanism then provides a reading on a dial of the speed of the shaft in r.p.m. The shaft is driven by a gear operated by a flexible drive from the engine. The drive may be at some fraction of the engine speed and suitable gearing in the indicator is employed to take this into account. The direction of rotation of the shaft is important, so a means is provided to ensure that this will be constant irrespective of the direction of rotation of the flexible drive.

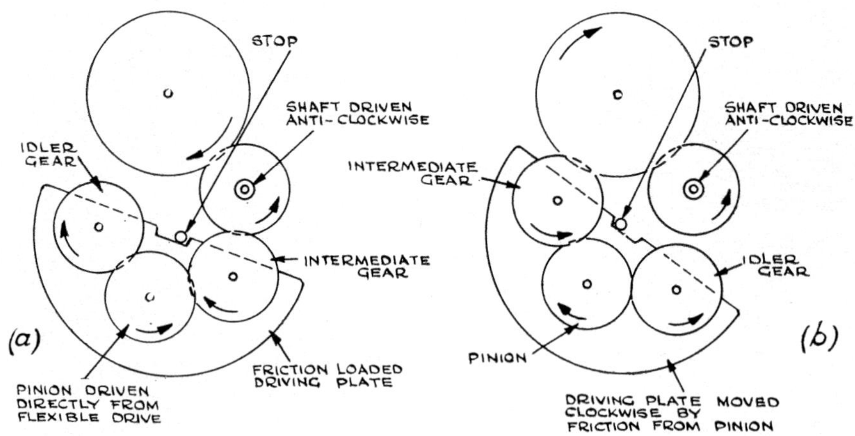

FIG. 8.3.

The mechanism used for this purpose is illustrated in Fig. 8.3 and comprises a driving plate which is frictionally loaded to the flexible drive and which moves against a stop according to the direction of rotation. In Fig. 8.3(*a*) it is assumed that the flexible drive is rotating anti-clockwise so that the right-hand side of the driving plate is against the stop. An intermediate gear carried by this plate moves into mesh with the pinion on which is mounted the shaft whose angular rotation is to be measured. In Fig. 8.3(*b*) the flexible drive rotation is clockwise so that the driving plate moves in the same direction until held by the stop. The intermediate gear that was meshing with the pinion is now idling out of mesh, while the gear that was previously idling is now meshing with another gear directly driving the pinion. In both (*a*) and (*b*) the direction of shaft rotation is the same.

Angular Velocity of Shaft. The shaft speed is proportional to engine speed and all that remains now is to convert this shaft speed into a reading in terms of r.p.m. This is done by means of a timing mechanism and an integrating device. The timing mechanism consists of an escape mechanism, similar to that of a watch, which is driven by the pinion previously mentioned, through a friction drive. The escape mechanism operates a series of cams at a time cycle which is constant, because of the friction drive, at nearly 1 second. One of these cams operates a spring arm in which is pivoted one end of the rotating shaft. The effect of this is to "rock" the shaft into and out of mesh with an integrating wheel. Since the pinion on the shaft is rotating at a speed dependent on the engine speed and is in mesh with the integrating wheel for a period of time governed by the timing mechanism, then the angular displacement of the integrating wheel is a measure of engine speed.

Integrating Mechanism. The integrating wheel is freely pivoted on a shaft and is restrained by a spiral spring. This means that while the rocking-shaft pinion is in mesh with the integrating wheel this will move forward against the pull of the spring. As the timing mechanism rocks the pinion out of mesh with the integrating wheel, so the wheel must return to its zero position under the influence of the spring. However, in moving forward, the integrating wheel moves a second gear, which, though mounted freely on the same shaft as the integrating wheel, has its own control spring. This wheel has a leaf spring which locates in the gear teeth to prevent its return as the integrating wheel is returned; the spring is operated by the cam device on the timing mechanism and only releases the wheel as the integrating wheel reaches its farthest point during the next cycle of operation. Thus, the second wheel is maintained in a fixed position for a given rotational speed of the shaft and it would be possible to mount the pointer directly on this wheel.

In fact, the meshing and de-meshing of the pinion with the integrating wheel may result in a one-tooth error that could cause slight oscillation of the pointer at a steady speed. To damp this oscillation, the second wheel is made to drive a stabilizing wheel via a spigot which operates in a clearance hole in the stabilizer. Thus the integrating wheel may cause the second wheel to oscillate, but this oscillation is not transmitted to the stabilizing wheel which operates the pointer. The complete mechanism is shown in Fig. 8.4.

Calibration. The calibration of the indicator is achieved by comparing it with a stroboscope and making necessary adjustments to the timing mechanism by increasing or decreasing the mass of a flywheel on the timing mechanism. In the chronometric instrument the slowing down of the escapement results in a longer meshing time between rocking pinion and integrating wheel and therefore an increased reading on the dial. Similarly, a reduction of mass speeds up the escapement and reduces the reading on the dial for a given shaft

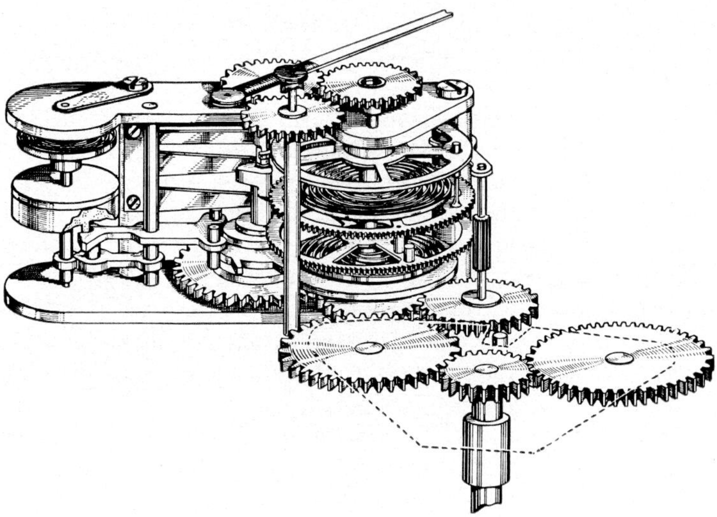

FIG. 8.4.

speed. The mass of the flywheel is varied by means of thin shims which can be added to or removed from the wheel. The instrument can be calibrated in this fashion to within a tolerance of ± 20 r.p.m. but manufacturer's literature should always be consulted when checking limits.

STROBOSCOPE

A stroboscope is an instrument for checking rotational speeds. The principle is very simple and may best be explained by the example shown in Fig. 8.5. Here, a ball is seen mounted at the end of an arm which is pivoted at the other end and made to rotate at 960 r.p.m. (16 revolutions per second). The cine-camera, shown at the right, is operated to take 16 pictures per second. Since the ball is also rotating at 16 revolutions per second, it will

will be realized that such a gyroscope fitted to an aircraft to define heading will show an error with time. Additionally, it will be seen that this wander would appear to indicate a clockwise movement to an observer at the North Pole but an anti-clockwise movement to an observer at the South Pole. Thus, in moving a gyroscope from the North Pole round the earth's surface to the South Pole the apparent wander would change from clockwise to anti-clockwise. At some point, therefore, there is no apparent wander, and this occurs at the equator. Thus, it is possible to plot apparent wander as shown in Fig. 9.6. It will be seen that the error follows a sine curve, and at any instant the error rate is 15 sin Lat. deg./hr.

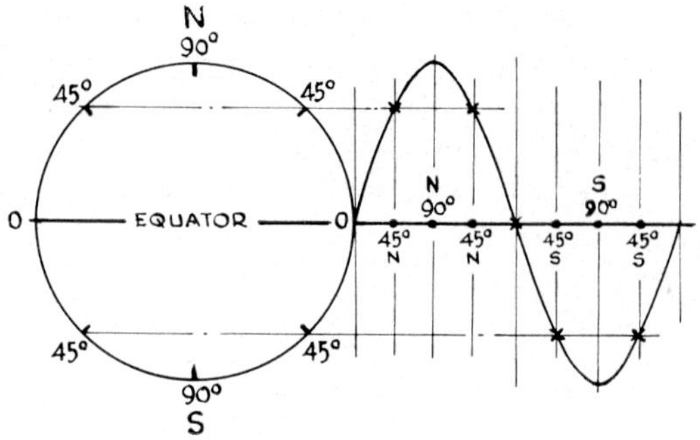

FIG. 9.6.

Correction for Wander. So far it has been assumed that the outer gimbal pivots are perfectly free and that no forces have been applied to the gyroscope; thus its inertia has kept it fixed in space. Now consider the application of a force on the gyroscope as shown in Fig. 9.7. The effect of this force is such as to tend to change the inner gimbal axis from *C–D* to C_1–D_1 and therefore the rotor axis from *A–B* to A_1–B_1; i.e. the force is felt at the inner gimbal pivots and through these to the rotor pivots and therefore at the edge of the rotor.

Consider the effect of this force on a small particle of the rotor which is travelling in a given direction at a particular velocity (Fig. 9.8). The applied

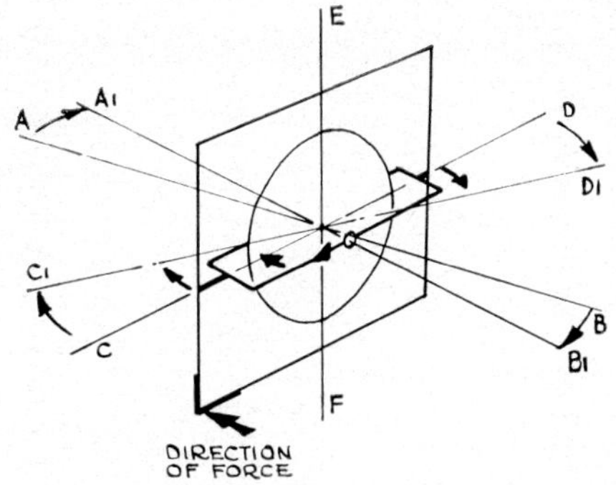

FIG. 9.7.

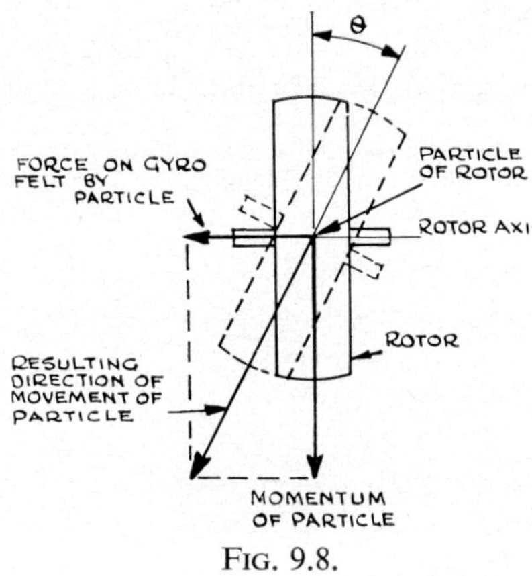

FIG. 9.8.

force on the outer gimbal ring is felt as a force on the rotor particle at right-angles to its direction of travel, as described earlier. The result is that the particle moves along a path which is the resultant of the two forces. If this is true of one particle it must be true of all particles of the rotor, which must therefore have moved through an angle θ. Hence, the application of a force on the gyroscope results in movement (known as precession) of the gyroscope at right-angles to the force.

It will be seen that if the mass and speed of the rotor is constant, the angle θ is proportional to the applied force. A sustained force will, of course, create a precession whose rate will be proportional to the sustained force. In the same way the application of a force on the inner gimbal axis will cause precession of the gyroscope about its vertical axis.

The angular velocity about the vertical axis is a measure of the sustained force applied to the inner gimbal. Thus, the sustained force of gravity on a weight suspended from the inner gimbal will produce a given rate of precession about the vertical axis. Here then is a means of correcting for wander. The inner gimbal is unbalanced so that gravity acting on the out-of-balance mass causes a precession rate equal to the rate of rotation of the earth. An observer would, therefore, detect no apparent movement and would declare no precession. Since the precession rate varies according to latitude it is not possible to compensate fully and so some wander will occur as an aircraft changes latitude.

In this description of wander, it has been assumed that the gyroscope was perfect in respect to balance, lack of friction and so on. In practice these also have an effect on wander, so that no calculations are required concerning the value of the mass needed to provide the compensation. Rather, this is simply a case of applying a small mass unbalance and checking the reading of the instrument after, say, one hour, and either increasing or decreasing the mass according to the change of reading.

Attitude Indication. So far the principles of the gyroscope have been applied to heading indication. Now consider the same gyroscope used for

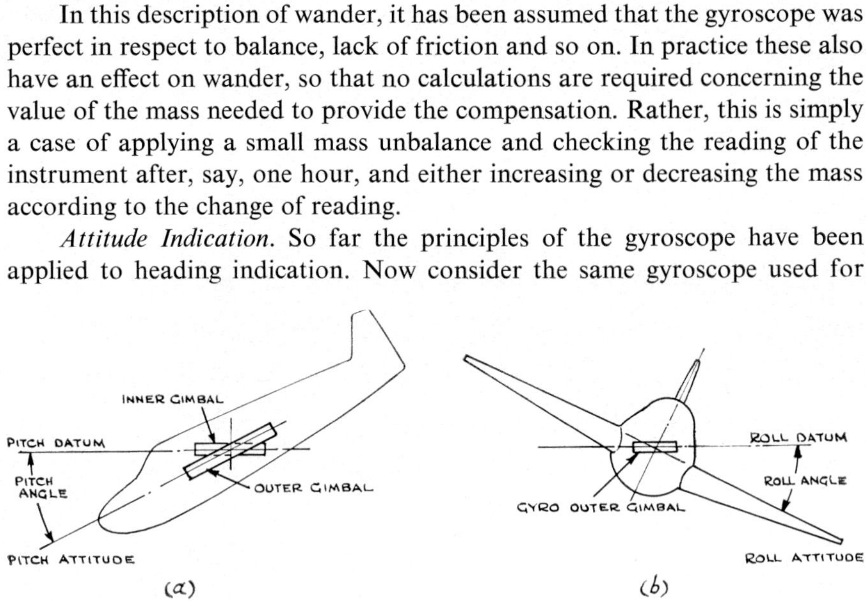

FIG. 9.9.

attitude indication, but with the outer gimbal pivots now lying in the fore-and-aft axis of the aircraft. This means that the spin axis of the gyroscope is vertical and any movement of the aircraft in pitch will show as an angular displacement between inner and outer gimbals (Fig. 9.9*a*). Similarly, a roll displacement of the aircraft will leave the gyroscope undisturbed, so that the roll angle will be indicated by the displacement between aircraft and gyroscope (Fig. 9.9*b*). The effect of wander in this case is such as to cause loss of accuracy of indication in pitch and/or roll. To provide the required datum for the aircraft the gyroscope spin axis must remain vertical with respect to the earth. A gyroscope of this kind is also provided with a gravity controlled device to ensure that this is so. These gravity devices are also subject to errors during turns, because of centrifugal force, but this problem will be discussed later.

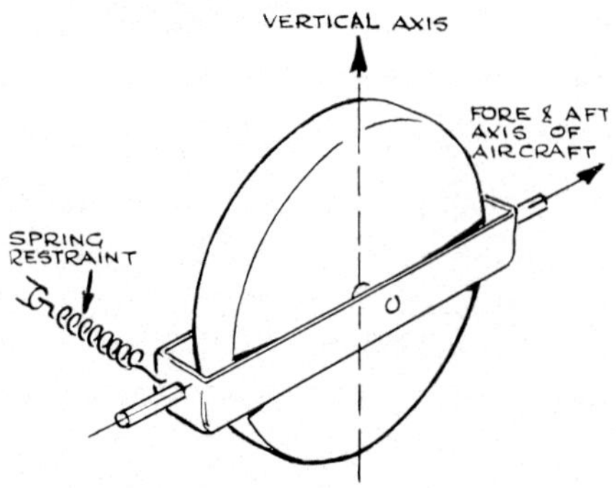

FIG. 9.10.

TURN INDICATOR

In this case the gyroscope is of a simpler type than that just described since it has one fewer gimbal ring. It consists of rotor and inner gimbal only, but the inner gimbal is spring-restrained (Fig. 9.10) and its pivots lie in the fore-and-aft axis of the aircraft. Should the aircraft turn about its vertical axis the inner gimbal pivots are carried with the aircraft and the resulting force is felt through the rotor pivots at the edge of the rotor (see page 109). The angular precession of the gyroscope rotor extends the spring until the spring tension balances the turning force. The angular displacement is therefore a measure of the turning force on the gyroscope. This varies with the rate of turn of the aircraft; hence, the angular deflection of the gyroscope is a measure of the rate of turn.

The turn indicator requires no datum with respect to gravity, and wander due to the earth's rotation can be ignored since the spring restraint allows the calibration to show, say, Rate 1 turns which are turns of 180 deg. (i.e. 180 deg./min.). The maximum wander rate uncorrected can only be 15 deg./hr. or $\frac{1}{4}$ deg./min. and the maximum error only $0{\cdot}25/180 \times 100 = 0{\cdot}139$ per cent.

PRACTICAL AIRCRAFT INSTRUMENTS

HAVING dealt with the principle of the gyroscope in so far as it concerns aircraft instruments, the constructional details of the finished instruments can be considered. The instruments to be described will be dealt with under the following headings: (1) direction indicator; (2) artificial horizon; (3) turn-and-slip indicator.

DIRECTION INDICATOR

It will be evident from the foregoing discussion that a gyroscope of the displacement type can be used as a reference from which changes of heading can be noted. It will be realized, however, that the direction indicator is merely duplicating the function of a normal aircraft compass. To realize the real need of the direction indicator, the limitations of the magnetic compass should be appreciated.

The simple magnetic compass consists of a pendulous magnet system that aligns itself with a north-south axis. Near the magnetic poles however the earth's magnetic field is such as to cause the aircraft magnetic compass needle to dip. The effect of this is to move the centre of gravity of the compass magnetic system away from its pivoting point. An aircraft heading north will thus have its magnetic needle down at the front and its centre of gravity behind the pivoting point. Should the aircraft now turn, the effect of centrifugal force on the centre of gravity causes the needle to turn in the direction of turn of the aircraft. Thus the magnetic compass gives an under-reading of the actual turn made. The error produced by this combination of circumstances is known as the Northerly Turning Error.

In effect then, the information given by the magnetic compass is perfectly reliable over long periods while the aircraft is flying straight and level, but is unreliable during the shorter periods during which manœuvres are being carried out. Contrast this with the principle of the gyroscope as previously discussed. The gyroscope was seen to provide a datum, but over a long period of time it showed a tendency to wander. Since it is hardly affected by centrifugal force it will provide an excellent reference over a short period of time.

The fitting of both a magnetic compass and a direction indicator to an aircraft means that the magnetic compass can be used to define the long-term heading, while the direction indicator can be used to define the short-term heading during manœuvres. The magnetic compass is used to monitor the direction indicator before the manœuvres are begun and the pilot then takes his reference from this until the aircraft settles down on a fixed heading. The magnetic compass is then used as the heading reference.

The mechanical direction indicator, or directional gyroscope, is of the displacement type described in the introduction and the rotor is made to spin by means of air jets. These jets are made to impinge on the rotor by an air pressure of $3\frac{1}{2}$ in. to 4 in. of mercury, or approximately 2 p.s.i. However, the pressure is derived from an engine-driven vacuum pump which is connected to the instrument case. This tends to evacuate the case, but a filter

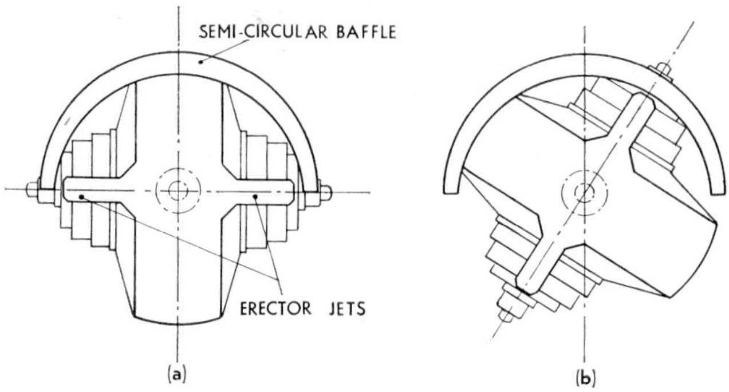

FIG. 9.11.

allows air to flow from the atmosphere to the rotor via channels cut into the gimbal system. Thus the air passing through the gyroscope is cleaner than if a pressure pump were used to blow air into the case.

Erection of Gyroscope. Since the gyroscope is dynamically balanced it may take up any position when it runs down and it is necessary to provide some means of erecting the gyroscope to a known datum. In one type of direction indicator this is done by arranging the gyroscope rotor in an inner gimbal ring which forms a box totally enclosing the rotor. The air driving the rotor must therefore find an outlet from this "box", which it does through two jets diametrically opposite. These jets react as a force on the gyroscope and therefore precession of the gyroscope occurs unless these two forces are equal, whereupon they cancel each other.

The two jets are arranged to be partially covered by a baffle plate (Fig. 9.11) when the gyroscope is erect, so that they are, in fact, equal. If the gyroscope is not erect, then one jet will be free of the baffle and will exert a greater force than the one still covered. Thus the differential torque from the jets provides for erection of the gyroscope.

Indication. By maintaining a pressure of $3\frac{1}{2}$ to 4 in. Hg, the gyroscope rotor will continue to run at a speed of up to 22,000 r.p.m. and the gyroscope will now be stable. Should the aircraft turn, the amount of turn will be indicated by the angle between gyroscope and aircraft. It is necessary to use this angular displacement to move a scale, or compass card, to indicate the angular change. This is done by the mechanism shown in Fig. 9.12. The direction indicator has two adjusting knobs at the front of the dial; one of these, the dial-setting knob, allows the instrument to be reset after any wander of the gyroscope. The other is the course-setting knob which is used to set a pointer to a desired course. Turning the aircraft on to the particular course will result in the card and course-pointer moving round until both appear against a fixed index. This indicates that the course has been reached.

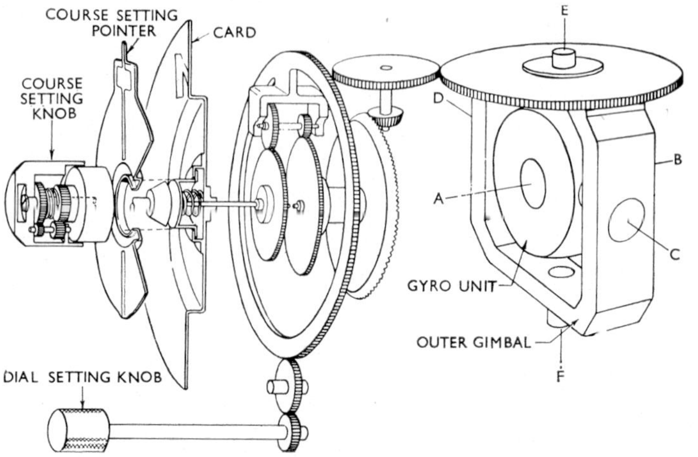

FIG. 9.12.

The maintenance of gyroscopes is rather specialized and it is essential to follow the manufacturer's instructions carefully. The obvious routine check should be made to ensure that the filter at the back of the case is functioning efficiently.

Testing. There are two main tests carried out on the direction indicator: (1) wander; (2) rotor freedom. The gyroscope is tested for *wander* by running it up to speed with the appropriate pressure applied. It is then set on to a particular heading and any wander from this heading over a period of time should be noted. The maximum wander rate should not exceed about ± 3 deg. after 15 minutes, but again consult the manufacturer's literature for exact values.

It should be noted that wander is caused not merely by the earth's rotation but also by inaccuracies and friction in the gyroscope bearings. For this reason the gyroscope is tested on a "table" which is capable of being moved in roll, pitch and yaw, so as to exercise the bearings. Such a table is the Bryans test table shown in Fig. 9.13. This comprises an electric motor geared to the table on which the gyroscope is mounted. A cranked lever allows the table to be tilted and the motor then provides the roll, pitch and yaw motion. A clock is built into the test table to enable the instrument to be timed. Provision is also made for automatic reversal of the motor and table action by means of a special switch in the clock mechanism.

The freedom of the rotor is checked by shutting off the air supply to it and noting the length of time it takes to run down. The time depends on the mass, speed and construction of the rotor and naturally varies with the design.

ARTIFICIAL HORIZON

This instrument is sometimes known as a *gyro horizon*, since a gyroscope is used to provide a reference intended to convey an idea of the earth's horizon. It is also referred to as a *gyro vertical*, since the indications given by the gyroscope are with respect to a vertical from the earth's surface. The *artificial horizon* provides pitch information by showing the vertical displacement of a model aircraft with respect to an artificial horizon represented by a pointer on the instrument dial. Similarly, roll information is provided by showing angular displacement between the model and the artificial-horizon pointer. The actual indications of the artificial horizon during various phases of flight are shown in Fig. 9.17. The air-driven artificial horizon is also a displacement gyroscope and is usually operated from the vacuum pump that drives the direction indicator, and therefore operates at the same pressure.

Erecting Mechanism. As with the direction indicator, it is necessary to provide an erecting mechanism, and again the idea of using evacuated air from the chamber containing the gyroscope rotor is used. It must be remembered that in this case the spin axis of the gyroscope is vertical and that it must be kept vertical with respect to the earth's surface. The erecting mechanism is therefore gravity controlled and consists of small pendulous vanes controlling four exhaust jets which provide the erecting torques (Fig. 9.14). This illustration also shows a heavy tungsten ball which prevents the gyroscope from tilting too far, as its rotor runs down, by bearing against the edge of the under side if the tilt angle tends to exceed 40 deg. Thus the gyroscope can be said to be "pendulous" when it is not running. It is not desirable that it should be pendulous when it is running, otherwise it would be affected by centrifugal force during a turn.

In Fig. 9.14 it will be seen that, with the gyroscope in the vertical position, the only effect of centrifugal force will be to displace the ball, and the gyroscope itself is not pendulous when running. The vanes themselves, being pendulous, will be affected by

FIG. 9.13. *Bryans test table, Mark 4a.*

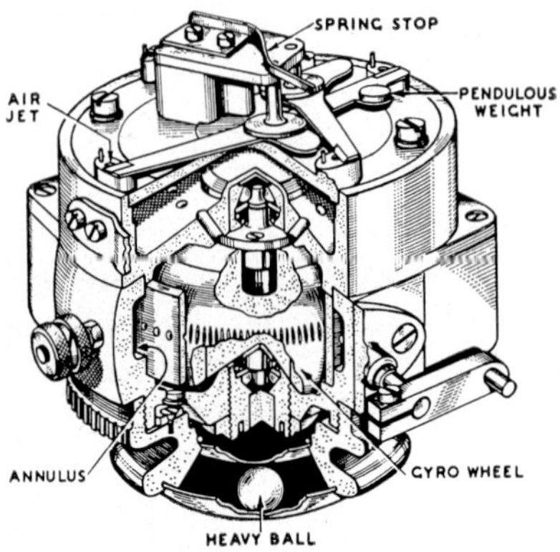

FIG. 9.14.

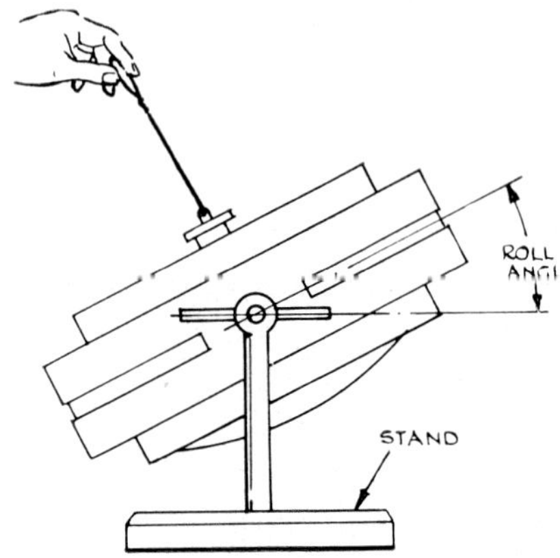

FIG. 9.15.

centrifugal force during a turn and this will create an error. The effect of this error is to reduce the indicated bank angle, but this bank error becomes a pitch error as the turn continues. It is difficult to show this diagrammatically, but a simple experiment will indicate that this is so; i.e. with an artificial horizon removed from the case and mounted in a stand, tie a thread to the inner gimbal and hold this so that the indicator shows a slight bank angle (Fig. 9.15).

Now hold the thread stationary and taut and rotate the stand through 90 deg., when it will be seen that the bank angle has disappeared and we are left with a pitch displacement. Moving the stand through a further 90 deg. will remove the pitch displacement and produce a bank angle of the same magnitude as the first bank angle but opposite in direction. This cyclic error, then, is a result of centrifugal force acting on the pendulous vanes during a turn. To compensate for this, it is arranged that the spin axis of the gyroscope is not exactly vertical but is tilted towards the nose of the aircraft (Fig. 9.16). It is held in this position by the vanes, which tilt the gyroscope until the vane platform is level with respect to the earth.

The effect of tilting the gyroscope axis is to produce bank and pitch errors during a turn, but the direction of these errors is opposite in sense to those produced by centrifugal force on the vanes. Thus, it is possible at some airspeed to compensate fully for turning errors.

Indication. The indication of pitch and roll is given by movement of a pointer with respect to a model aircraft. This pointer is in the form of a horizontal bar placed across the instrument dial, the bar being regarded as an "artificial horizon" (from which the instrument gets its name). Pitch indication is given by the horizon bar or pointer moving up or down, while roll indication is given by angular movement of the horizon bar. A pointer at right-angles to the horizon bar indicates the roll angle (Fig. 9.17).

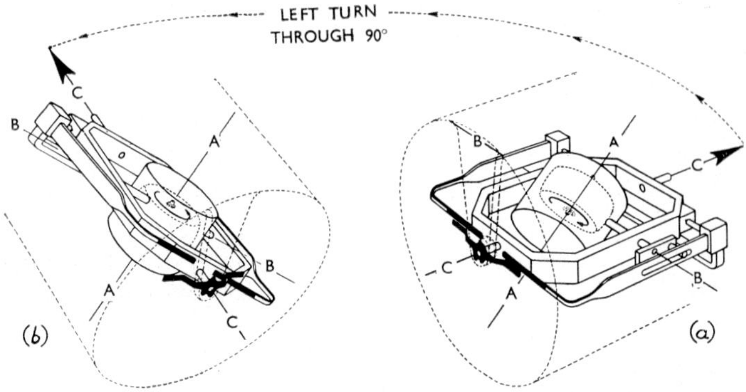

FIG. 9.16.

Maintenance and Testing. This instrument requires special maintenance and it is essential to see that the air filter is kept clean. Testing must be done according to the manufacturer's specification but, in general, consists of tests (i) to (iv).

(i) *Erection Test*. This test is to ensure that the instrument is erect and ready for use soon after starting up the engines which drive the vacuum pumps. The erection should be complete in about $1\frac{1}{2}$ minutes.

(ii) *Erection after Toppling*. The gyroscope is now "toppled"; i.e. it is moved so that the limit of the freedom of the gyroscope is exceeded, thus causing precession. Once the position is restored it will re-erect itself and

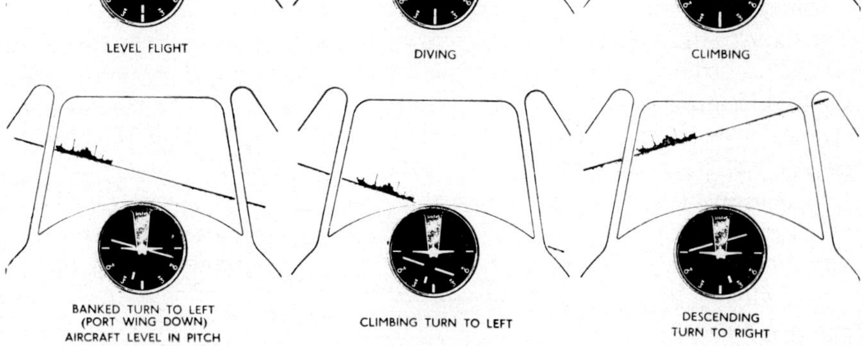

FIG. 9.17.

the time taken for this is noted. The manufacturer's literature will quote appropriate re-erection times.

(iii) *Roll, Pitch and Yaw Test.* In this test the gyroscope is mounted on a test table, such as the Bryans Mk. 3, and a roll, pitch and yaw motion is induced. After about 10 minutes' operation, the table is restored to its horizontal position, whereupon the indication should show no change from the original.

(iv) *Rotor Freedom.* Shutting off the air supply, note the time taken for the rotor to coast to rest. This time varies according to the design.

TURN-AND-SLIP INDICATOR

The turn-and-slip indicator really comprises two separate indicators in one instrument case. The turn indicator is a gyroscope of the *rate* type which indicates the rate of turn of the aircraft about its vertical axis. At first sight it may be thought that the turn indicator is merely repeating the information of the direction indicator, which will also tell the pilot if he is turning the aircraft from a fixed heading. However, if the aircraft is turning away from course at say 30 deg./minute the direction indicator must take one minute to show the 30 deg. displacement. The turn indicator on the other hand will respond to the 30 deg./minute turn immediately and the pilot is made aware of the fact that the aircraft is leaving a fixed heading datum. The slip indicator is used in conjunction with the turn indicator to enable the pilot to co-ordinate the aircraft turn when manoeuvring.

Mechanical Details: Turn Indicator. This is a much simpler instrument than the direction indicator or artificial horizon in that there is no necessity for erecting devices or wander corrections. The turn indicator consists of a gyroscope of the rate type; i.e. one comprising rotor and inner gimbal only, the gimbal being spring-restrained. The rotor is driven by air jets, as in the previous gyroscopic instruments, but the speed of the rotor is very much lower since the degree of stability is not so high.

Because the gyroscope is arranged with its inner gimbal axis lying in the fore and aft axis of the aircraft, a turning movement about the vertical axis will produce a torque on the gyroscope and cause precession. The value of the torque is a measure of the "rate of turn" about the vertical axis, and the precession increases until the extended spring tension balances this torque. Thus, the angular deflection of the gyroscope is a measure of the rate of turn. A pointer operated by the inner gimbal moves over a scale calibrated in units which represent a particular angular velocity; e.g. the first unit on the dial represents a turn of 180 deg./minute.

Damping. Because of the spring-restraint, a torque on the gyroscope would produce an oscillation of the pointer before giving a steady reading. A damping device is therefore incorporated to make the pointer "dead-beat".

This damping device (Fig. 9.18) consists of a piston which is attached to the inner gimbal and which works in a cylinder. The clearance between piston and cylinder is very small, so that an air pressure builds up as the piston moves into the cylinder. A small orifice allows air to leak slowly away; thus, oscillation of the gyroscope is cushioned by this air-dashpot.

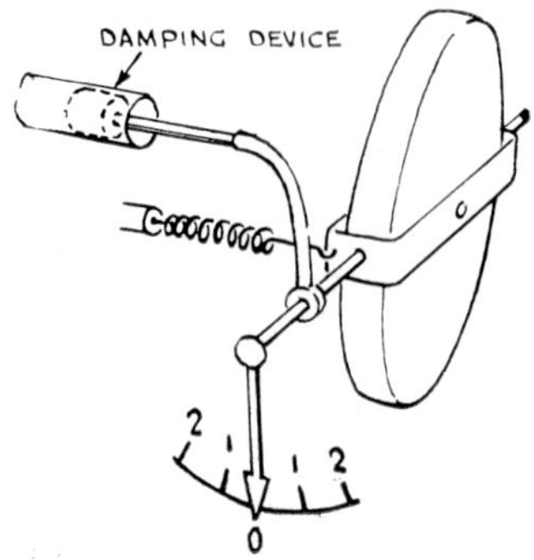

FIG. 9.18.

Maintenance and Testing. Again, in this instance, the manufacturer's literature should be consulted, but the air filter can be given the usual attention. Testing comprises items (i) and (ii), which follow.

(i) *Calibration Test*. The turn indicator is tested on a turntable, such as the Bryans Mk. 3, by rotating the table at such a speed as to produce, say, a Rate 1 turn. The speed of the rotating table can then be ascertained and instrument errors noted. Adjustment for errors consists of adjustment of the spring-restraint on the inner gimbal. An increase of spring tension will produce a reduction of indicator reading for a given rate of turn. (If the instrument is mounted in the normal position on the table it will not be simulating aircraft conditions, since the aircraft is banked at an angle during a turn. It is usual, therefore, to mount the indicator in a frame which allows it to be fixed at an angle before the table is rotated. Details of the angle at which to tilt the gyroscope for a given rate of turn can be obtained from manufacturer's literature.)

(ii) *Rotor Test*. The freedom of the rotor can be checked by shutting off the air supply and noting the time taken for the rotor to stop.

SLIP INDICATOR

This is the second half of the turn-and-slip instrument but is really a separate instrument in its own right. However, because it is usual to try to

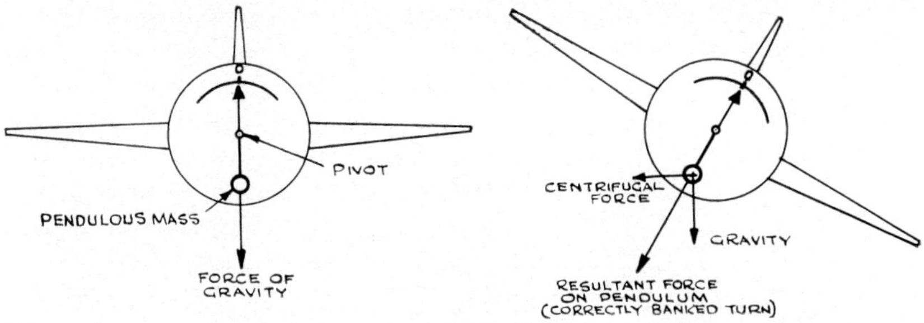

FIG. 9.19.

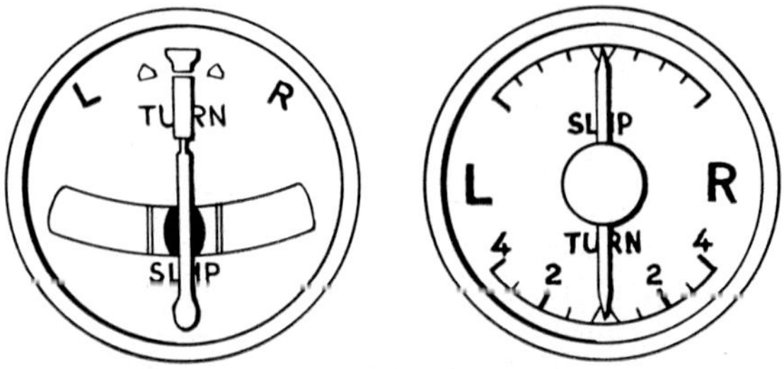

FIG. 9.20.

turn an aircraft without sideslipping into or out of the turn, the slip indicator is mounted in the same case as the turn indicator so that both indicators can be seen at once.

The slip indicator usually consists of a pendulum which operates on to a pointer. During normal flight, the effect of gravity is such as to hold the pendulum in a position where the pointer will indicate zero. Should the aircraft turn, the effect of centrifugal force on the pendulum will cause it to take up a position which is the resultant of centrifugal and gravitational forces. The position of the pointer will still be zero for a correctly banked turn (Fig.9.19). In order to produce "dead-beat" indications it is usual to provide some form of damping for the pendulum. This can take the form of air-dashpot damping similar to that employed for the turn indicator.

Ball-type Slip Indicator. A simpler type of slip indicator is the ball type. This consists of a curved glass tube which contains an agate ball. As in the pendulum type, the ball will remain central so long as a correctly banked turn is executed. Sideslip is shown by the ball moving away from the zero position. In this instance damping is provided by a liquid which almost completely fills the tube in which the ball operates. A small air space is left in the tube to allow for expansion of the liquid. It is not customary, however, to see any air bubbles in the liquid, but if they appear it is necessary to remove the instrument from the panel and rotate it about its horizontal axis. This allows the trapped air in the tube to enter an expansion chamber concealed behind the instrument dial. The testing of either the pointer or ball-type slip indicator is fairly simple: (i) Place instrument on a flat-plate and note that ball or pointer is in the centre of the scale. (ii) Rotate the instrument about its horizontal axis and note that the pointer or ball moves smoothly to and fro and returns to the zero position.

Instrument Presentation. The presentation of the instrument naturally depends on the type of slip indicator. In the case of the pendulum type of slip indicator the instrument shows two pointers, the upper one indicating *slip* and the lower one *turn*. When the ball type of slip indicator is used, the slip indicator occupies the lower portion of the dial, while the turn indicator has its scale at the top of the dial (see Fig. 9.20). Usually there is no calibrated scale, but a mark either side of zero is normally equivalent to a Rate 1 turn; i.e. a turn of 180 deg./minute.

Part II

ELECTRICAL INSTRUMENTS

CHAPTER 10

Remote Indication

BECAUSE of the growing size of aircraft and the attendant difficulties of continuing the use of mechanical instruments (owing to the long pipe-lines, flexible drives and so on involved) a gradual change towards the use of electrical instruments has taken place over the past few years. These instruments usually consist of two separate elements; a transmitter and a receiver. The receiver is wired electrically to the transmitter to form a complete system. The transmitter is a device which measures a change in some variable and supplies information to the receiver which indicates the change.

For ease of maintenance, supply of spares and so on it is of advantage to use the same type of system for measuring different variables, and many such systems are available. They can however be reduced to two classes of system: d.c.-operated systems; a.c.-operated systems. To illustrate the underlying principles of these systems two British systems have been chosen: the Desynn system (a d.c. synchronous system); the Aysynn system (an a.c. synchronous system).

DESYNN SYSTEM

The Desynn system is a simple but effective electrical method of remote indication designed for continuous reading on a 12- or 24-volt d.c. supply. (The name Desynn implies a d.c. synchronous system.) It consists of two main parts: a transmitter, and a receiver (or indicator). The transmitter consists of a toroidal resistor (Fig. 10.1) mounted in a suitable moulding which also acts as a terminal block, having five shrouded terminals. Three tappings are taken from the resistor at intervals of 120 deg. and are soldered to Nos. 1, 2 and 3 terminals on the moulding. Two light phosphor-bronze brushes rest on the resistor at diametrically opposite points. One brush is in direct contact

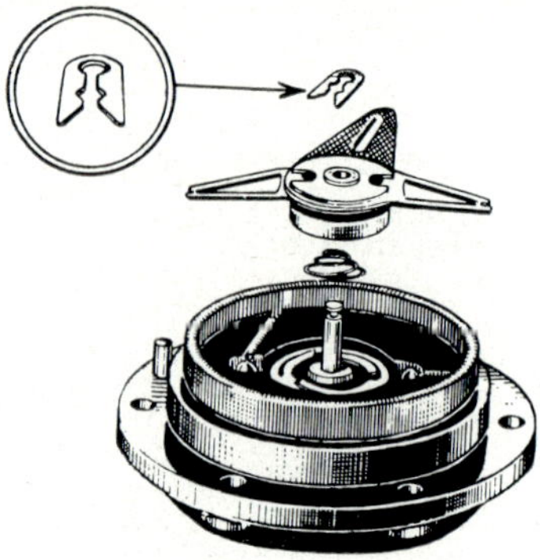

FIG. 10.1.

with a central spindle about which the brushes can rotate. This spindle also acts as a terminal, being connected to the positive side of the supply voltage. The other brush is in contact with the negative terminal by means of a slip-ring and a spring loaded ball bearing, or, in later models, a light three-fingered spring. The brushes are insulated one from the other and are capable of continuous rotation. They are always diametrically opposite and, being connected across a supply voltage, feed in a current to the toroid, the value and direction of current depending on the position of the brushes on the resistor.

The indicator (Fig. 10.2) consists of a rotor *C* and a stator winding *F*. The stator winding is of the three-phase star-connection type consisting of six coils, these being held in position by a band or yoke of mumetal which also forms part of the magnetic circuit. The rotor is magnetized and fitted on to a spindle which is freely pivoted in a brass housing *D*. A hardened steel screw *E* provides for end-play adjustment of the pivots in the region of 0·003 to 0·005 in. A pointer is fitted to the rotor spindle and is capable of 360 deg. rotation. In general practice, however, a scale of only 300 deg. or 180 deg. is used, and the pointer is arranged to move off the scale when the current through the indicator is interrupted. This is done by arranging a weak fixed magnet *A* in such a position that the rotor magnet lines up with it to carry the pointer off the scale. Other features indicated in Fig. 10.2 are the end plate *B* and the end cover *G*.

Three leads, one from each winding, are taken to a terminal block at the rear of the instrument and are numbered 1, 2 and 3; these terminals are connected to terminals with like numbers on the transmitters. The circuit is shown in Fig. 10.3. It will be seen that a current will flow through the transmitter to the indicator windings, the magnitude and direction of current being dependent on the position of the brushes on the resistor. A magnetic field is set up (in the indicator coils) the direction of which will change with changes of position of the brushes on the resistor. It will be found that a

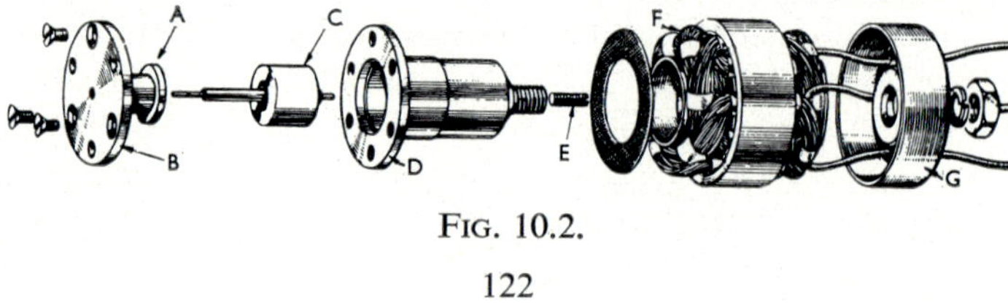

FIG. 10.2.

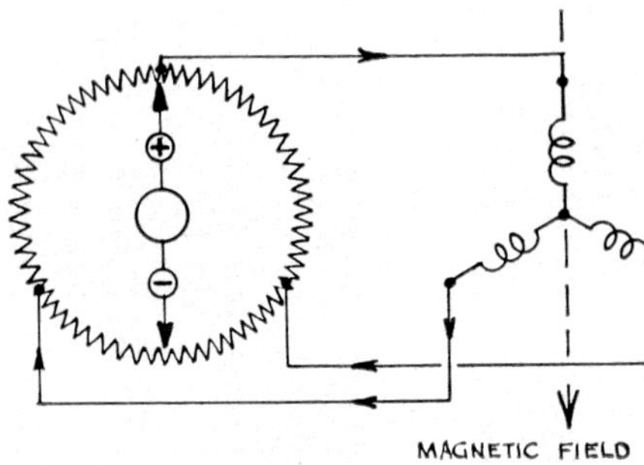

FIG. 10.3.

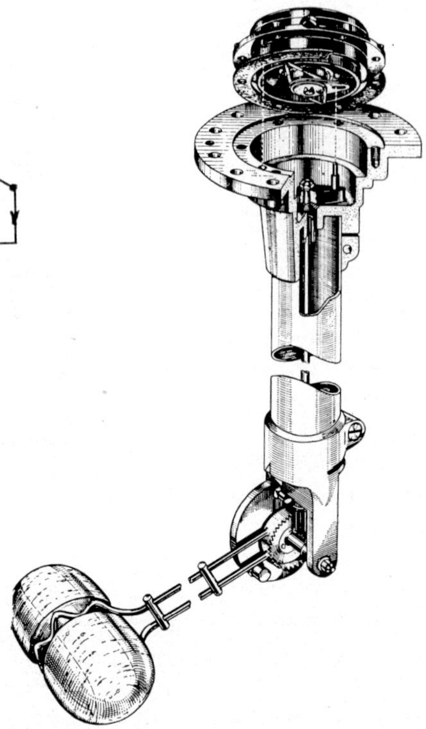

FIG. 10.4. *Fuel-contents gauge tank unit.*

rotation of the brushes through a known angle will produce a rotation of the field through a like number of degrees. The rotor aligns itself with the magnetic field and carries the pointer to a position which is an indication of the amount of brush movement.

In practice, the pointer movement is not quite equal to the brush movement, since the pull-off-magnet, if fitted in the indicator to give an off-scale pointer reading, slightly alters the scale shape. The pull-off magnet always acts in such a way as to oppose the movement of the rotor through the first half of the scale and assist the movement through the other half. It must be made clear, however, that this only results in a slightly uneven calibration of the scale, and in no way detracts from the efficiency of the system.

If it is desired to have an evenly calibrated scale, so matching pointer movement to brush movement, the pull-off device is omitted. This means, however, that when the circuit is switched off the pointer will stay on the scale at its last indicating position. When the circuit is switched on again it will go to its new indicating position by the shortest path, not necessarily through zero.

The Desynn system has innumerable uses but is perhaps best known as a fuel-contents gauge, as a position indicator (flaps, gills, etc.) and as a pressure gauge. While in each case the same type of indicator is used (with appropriate calibration) the transmitter and its method of operation will depend on the use to which the instrument is put, as shown in (i), (ii) and (iii).

(i) *Fuel-contents Gauge.* This consists of the transmitter and indicator already described and a float assembly to operate the transmitter. The float assembly consists of a cork float (suitably treated so as to be impervious to liquids) and an arm which operates a system of gears and, through a spindle and crank lever, the electrical transmitter (see Fig. 10.4). The float assembly is designed for a particular shape and size of tank and is interchangeable only in tanks of the same kind. Similarly, the indicator is calibrated for a particular kind of tank and should not be used for any other type. Stops are fitted to limit the travel of the float arm.

FIG. 10.5.

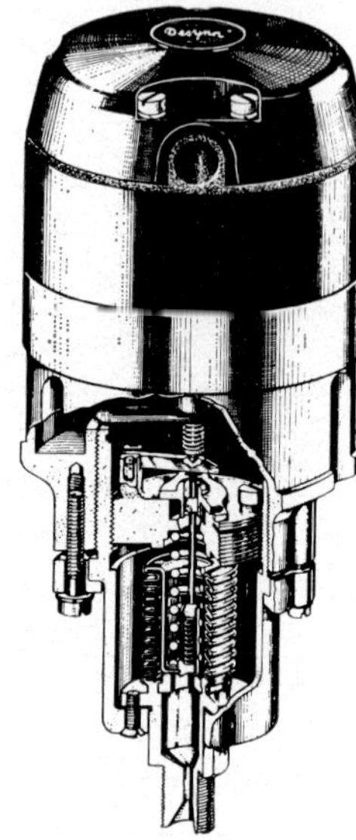

FIG. 10.6.

(ii) *Position Indicators.* To indicate flap or gill positions a standard indicator and electrical transmitter are used, but the method of operation of the transmitter is slightly different from that employed on the fuel-contents gauge in that the transmitter housing contains a crank lever which is operated by an adjustable lever arm through a system of gears (Fig. 10.5). The lever arm is normally operated through 60 deg. and the gear ratio in the housing is 3 : 1 or 6 : 1 so that the transmitter brushes (and therefore the indicator pointer) move through 180 deg. or 360 deg. Stops limit the movement of the lever arm to 70 deg.

(iii) *Pressure Gauges.* To enable the Desynn system to be used for indicating pressure, a "micro-transmitter" has been developed. This is used in conjunction with a suitable calibrated standard indicator. It consists of two parts: pressure unit; electrical transmitter.

Pressure Unit. This consists of a small chamber which can be connected directly to the supply pressure to be measured. The chamber is formed by a bellows unit (Fig. 10.6) soldered around the base to a screwed metal connection and restrained by a spring against a top plate. The spring strength varies according to the range of pressures to be measured, typical ranges being:

Oil: 0–120 or 200 lb.
Fuel: 0–10, 20, 30 or 40 lb.

When pressure is applied to this chamber the bellows move against the tension of the spring, this movement being conveyed by means of an operating pin (which passes through the top plate) to a lever pivoted on the top plate. Whatever the range of this unit, the movement transmitted to the lever is limited to 0·1 in. between zero and maximum pressure. The reason for this standard movement will be seen later.

Electrical Transmitter. This is electrically the same unit as the toroidal type of transmitter, but is operated by a pin-and-lever device against the tension of two light coiled springs. The brushes in this instance travel through approximately 60 deg. and bear on two resistance bobbins (Fig. 10.7) which lie side by side. The method of interconnection of the resistors, however, gives the same results as a 360 deg. movement of the brushes on a toroidal resistance. This can be explained in the following way. Imagine two toroidal resistors mounted so as to form parallel circuits with one brush moving over each resistor, the connection being as shown in Fig. 10.8. This would be found to function electrically in the same way as a single toroid. Suppose the outer of these two resistors to be broken at tapping point 3, and opened out to form a straight line, points 3-3 then being connected together. The same procedure is now adopted with the inner resistor, this time breaking it at a point midway between tappings 1 and 2 and short-circuiting the ends of the resistor.

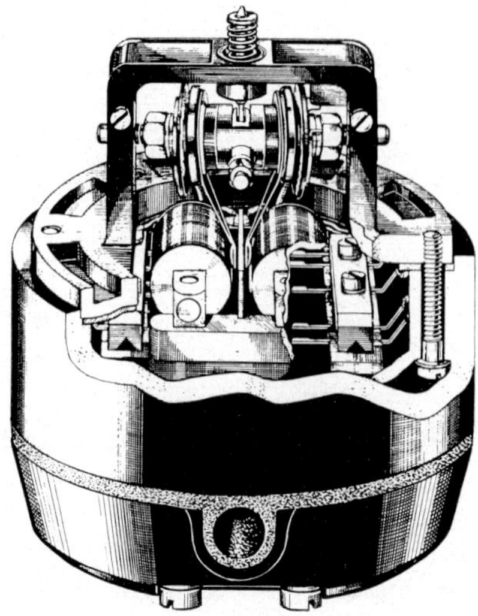

FIG. 10.7.

The two resistors are connected side by side as in Fig. 10.9, from which it will be seen that they provide the same circuit as the double toroid. The supply is connected across two brushes which move in unison from *A* to *B* when the pin-and-lever device is operated. The transmitter pin is set up in

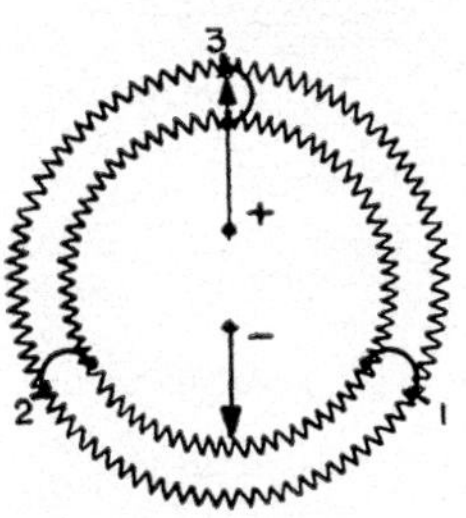

FIG. 10.8.

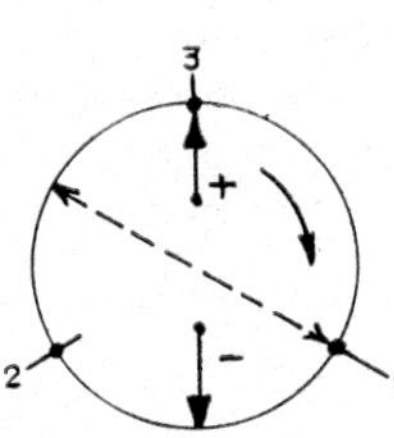

FIG. 10.9.

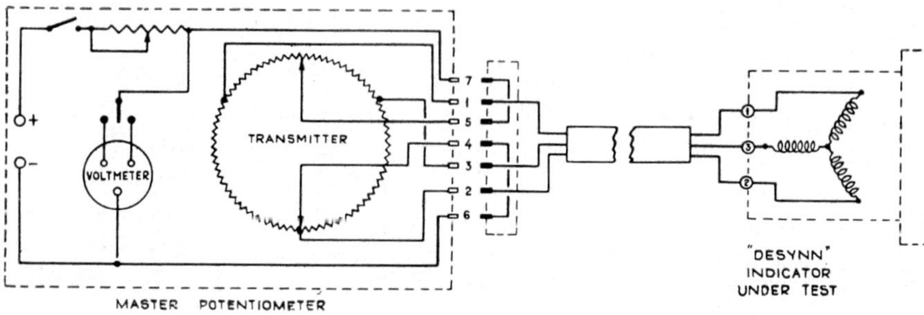

FIG. 10.10.

such a way that a movement of 0·10 in. is sufficient to cause full travel of the brushes over the resistors. When transmitter and pressure unit are connected together, the transmitter pin rests on the pressure unit lever and acts, as it were, as an extension of the pressure-unit operating pin. Therefore, a pressure covering the full range of the unit will give full travel of the brushes over the resistors. Hence, a Desynn indicator can be used to show a suitable range of pressures, the pointer covering as it does, 300 deg. of the scale.

Testing the System. The testing of Desynn indicators and transmitters is carried out in two separate parts. Indicators are checked against a Desynn master potentiometer (Fig. 10.10) which is calibrated in mechanical degrees. A calibration chart for the particular instrument gives degrees as shown on the master potentiometer for an equivalent reading on the indicator

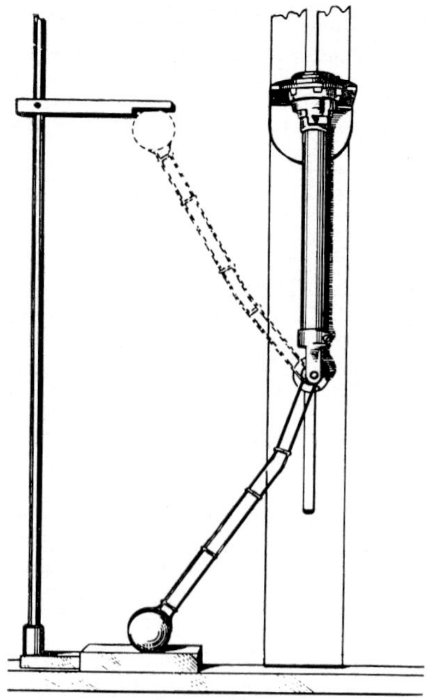

FIG. 10.11.

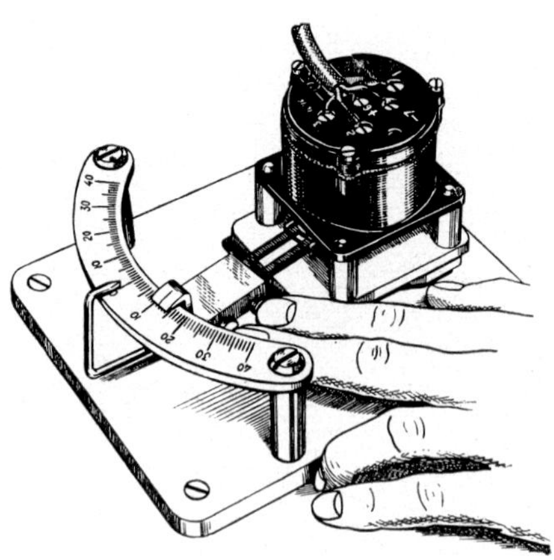

FIG. 10.12.

under test the tolerance being of the order of ±5 deg. In the case of the fuel gauge no positive errors are allowed at the lower end of the scale, and elsewhere the tolerance is limited to ±4 deg.

Transmitters are checked against a master Desynn indicator using some form of jig or fixture to hold the transmitter. Typical jigs are described in the following paragraphs. In the case of tank units only the top and bottom errors can be ascertained and the tolerances permitted are of the order of −3 deg. at the bottom and ±4 deg. at the top of the scale (see Fig. 10.11). (The lower tolerance does not permit a positive error since a fuel gauge is not allowed to over-read for safety reasons.) The position transmitter is connected to a fixture of the type illustrated in Fig. 10.12 and the lever arm is moved to a mid-position. The Desynn master indicator should now read 180 deg. Moving the arm either way through 30 deg. should result in a master-indicator change of 90 deg. or 180 deg. according to whether the transmitter is a 3 : 1 or 6 : 1 ratio type. A tolerance of ±4 deg. is allowed in either case. The pressure transmitter is tested, as for normal pressure gauges, by connecting it to a dead-weight test rig (Fig. 10.13). The transmitter is then electrically connected to the master indicator and pressure gradually applied via the dead-weight tester. At each succeeding higher pressure, the master-indicator reading is taken, and a chart (supplied by the manufacturers) will show the error of the transmitter at any point. The tolerance is of the order ±3 deg. at zero and ±7 deg. over the rest of the scale.

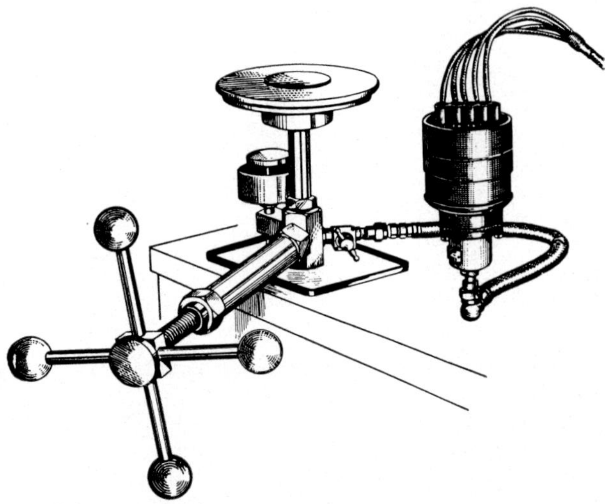

FIG. 10.13.

SLAB-DESYNN

A further application of the system is in the so-called "Slab-Desynn" transmitter and indicator. At the outset it must be made clear that although transmitter and indicator function on the Desynn principle they are not interchangeable separately with the transmitters and indicators previously described, neither may the standard 24-volt master potentiometer and indicator be

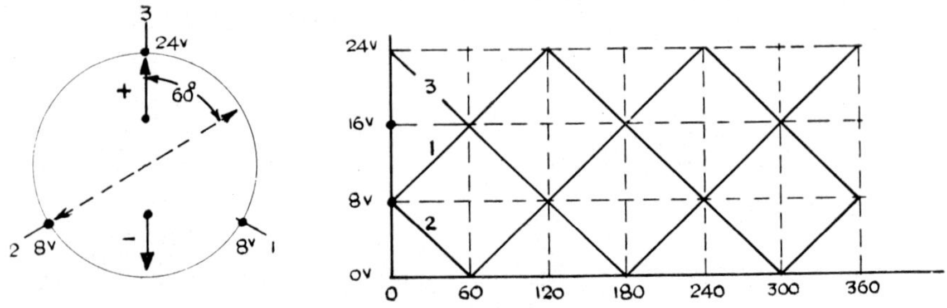

FIG. 10.14.

used for testing slab-type transmitters or the indicators used with them.

The purpose of the Slab-Desynn system is to provide more accurate indication than can be given with the conventional Desynn system. The Slab-Desynn overcomes two of the sources of error found in the normal Desynn system, and caused by friction and by the waveform. In the conventional Desynn transmitter there is a considerable surface of resistance wire over which the brushes must move, and this gives rise to small errors owing to the friction between the brush and the resistor. An error that is not so obvious appears in the Desynn system as a result of the output waveform of the normal transmitter. This waveform can be shown as represented in Fig. 10.14. This shows that, with the brushes in the position given, the potential at the three tapping points will be as at the zero position on the graph; i.e. tapping 3 is at maximum volts, while tappings 1 and 2 are each at 8 volts (Fig. 10.15). These are the voltages applied to the indicator in order to cause currents to flow through the indicator coils. The resulting field is accordingly governed in direction and magnitude by the voltages applied.

Should the brushes be moved clockwise through 60 deg. the voltages across the indicator coils will then be as shown in Fig. 10.16. The currents through the indicator will now produce magnetic fields whose resultant will be displaced 60 deg. from the previous resultant. Thus the indicator field lines up exactly with the brush movement. However, if voltage values are taken

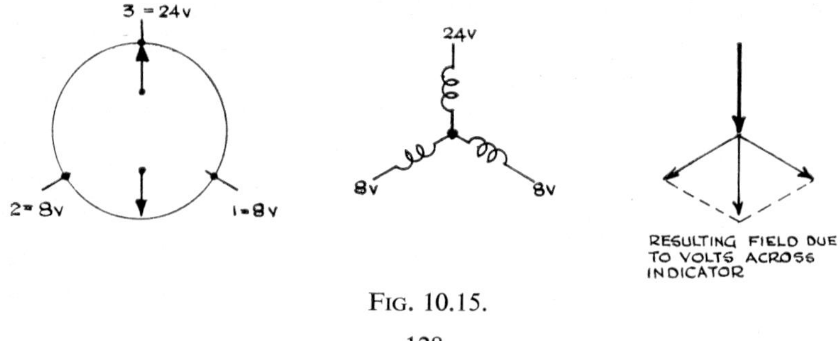

FIG. 10.15.

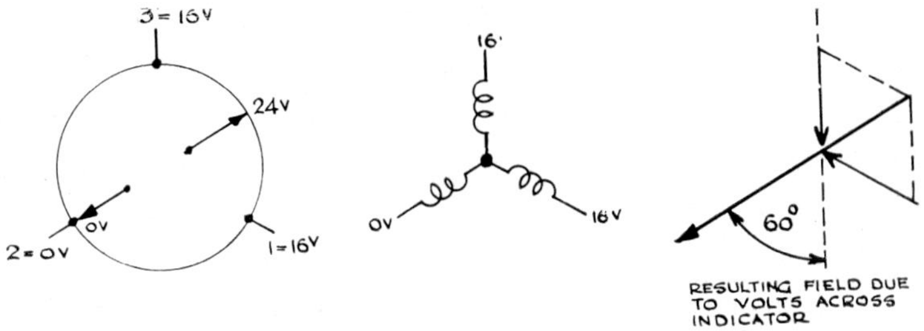

FIG. 10.16.

from the graph at 7½ deg., 22½ deg., etc. and the resultant field is worked out it will be seen that it does not agree with brush movement and that a cyclic error exists which reaches a maximum every 15 deg.

It will be noted that the mean value of the voltage of the graph is 12 volts and that the contact voltages are varying about this mean value with a saw-tooth wave. Now if the wave shape could be changed to one of a sinusoidal form the cyclic error would disappear and the field position in the indicator would agree exactly with the brush position of the transmitter at any point. The Slab-Desynn transmitter achieves this waveform and at the same time reduces friction by having a smaller contact surface between brush and resistor, the latter being highly polished.

This transmitter comprises a slab former on which is wound a resistance wire shown schematically in Fig. 10.17. The electrical supply is connected across the resistor, and three brushes (insulated from each other) are pivoted on the centre of the slab. Rotation of the brushes produces a sinusoidal wave-form to reduce the cyclic errors present in the toroidal and micro-type trans-mitters. It should be noted, however, that this waveform assumes that there is an open circuit and therefore when the indicator coils are connected to the brushes the open circuit no longer exists. There is thus a tendency to spoil the sinusoidal waveform, but this is reduced to a minimum by arranging that

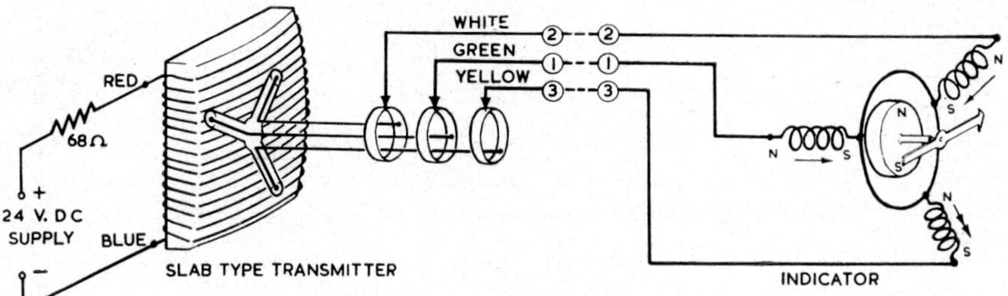

WHEN THE TANK UNIT IS INVERSELY MOUNTED THE GREEN TRANSMITTER LEAD IS CONNECTED TO TERMINAL ③ AND THE YELLOW LEAD TO TERMINAL ①

FIG. 10.17.

9—AI

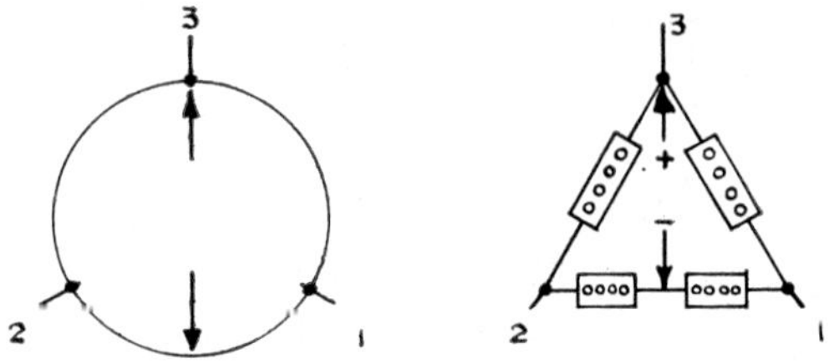

FIG. 10.18.

the resistance of the indicator is high enough to approximate to an open circuit. As mentioned earlier, this means that the Slab-Desynn transmitter and the indicators used with it are not of standard form and should not be checked with the normal Desynn test sets.

TESTING AND USE OF SLAB-DESYNN EQUIPMENT

To test the operation of the Slab-Desynn transmitter a master Desynn indicator is used. While this is identical in appearance to the ordinary Desynn master indicators, its resistance values are different, and it should not be used to check ordinary transmitters. The transmitter to be checked is connected to the correct master indicator and checked in accordance with the manufacturer's instructions.

Indicators are checked by making use of standard decade resistance boxes. The principle of these is simple; e.g. using a toroidal transmitter to illustrate the principle, it will be seen that the two electrical circuits shown in Fig. 10.18 are identical. The decade resistance values are set to equal the value of resistance between the respective tappings and brushes. Similarly, it is possible to produce the equivalent of any transmitter position. The circuit for testing indicators is shown in Fig. 10.19.

The main uses of the Slab-Desynn are: (*a*) as a torque pressure system; (*b*) as a fuel-contents gauge. These are described in the following paragraphs.

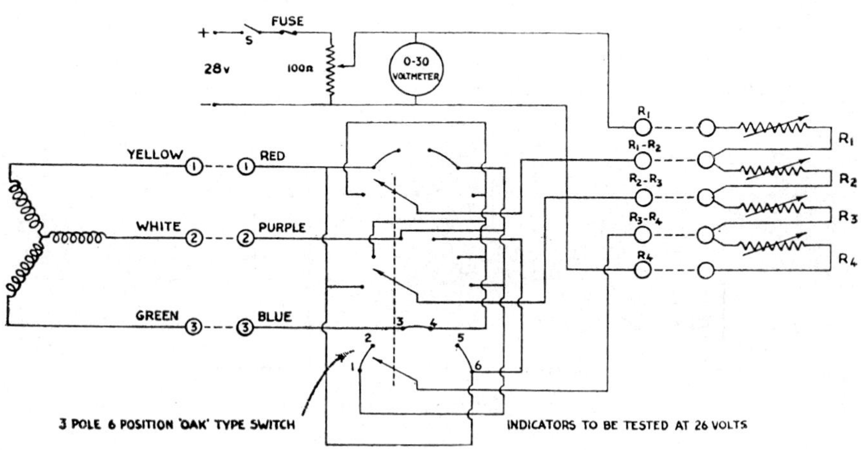

FIG. 10.19.

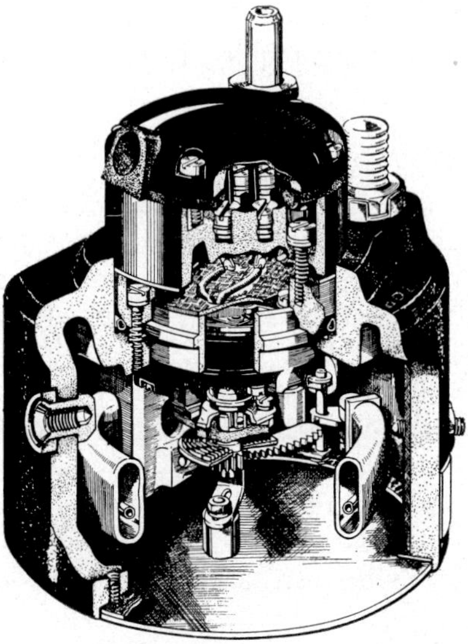

FIG. 10.20. *Slab-Desynn transmitter and hydraulic pressure unit.*

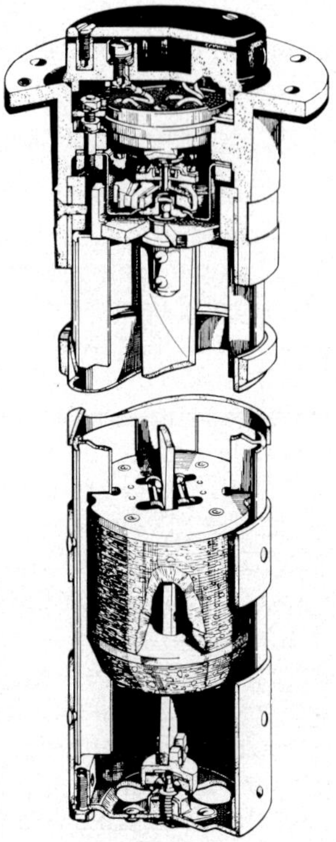

FIG. 10.21.

Torque Pressure System. This is a straightforward measurement of an oil pressure which is proportional to engine torque (Fig. 10.20). The indicator is usually calibrated in lb. per sq. in. The tolerance allowed on this system is ± 2 deg. at the indicator and ± 3 deg. at the transmitter. It will be seen that this is very different from the standard Desynn pressure system where the tolerances were shown to be ± 5 deg. and ± 7 deg. for indicator and transmitter respectively.

Fuel-contents Gauge. This consists of a tubular tank unit (Fig. 10.21) containing a float which rises or falls with fuel level. The float causes an arm to rotate in the tank unit thereby operating a Slab-Desynn transmitter through a magnetic drive. The indicator and transmitter are tested, as already outlined, by means of decade boxes for the indicator, and a master indicator for the transmitter.

Colour Codes. It is worth noting that the Desynn systems have been colour-coded as follows:

Voltage		*Colour Code*
12	Normal system	Green
24	Normal system	Yellow
24	Slab system	Purple

Both "green" and "yellow" indicators can be tested on the Desynn master transmitter, while the "purple" indicators are checked with decade boxes. In the case of transmitter checks, each system requires its own master indicator.

AYSYNN SYSTEM

The Aysynn system can be described as an "a.c. synchronous system" of transmission. This system is used to form part of many separate indicators, for example radio magnetic indicators, radial magnetic selectors, compass systems and integrating flowmeters all use Aysynn-type systems as part of the complete indicator.

The Aysynn system also has a transmitter and indicator (or receiver). Both transmitter and receiver consist of a star-wound stator with a wound two-pole rotor. The two stators (Fig. 10.22) are connected to one another and both rotors are connected in parallel across a single-phase a.c. supply usually of 26 volts, 400 cycles per second.

Considering the transmitter alone it will be seen that this is really a transformer in which the energized primary (the rotor) can be moved relative to three secondary windings which are displaced from each other by 120 deg. An e.m.f. is induced in each leg of the stator, the value being proportional to the flux linkage, which in turn depends on the angle between the particular secondary and the primary. In Fig. 10.22 the whole of the flux from the primary links with coil 3 to produce maximum e.m.f. but, because of the angle between primary and coils 1 and 2, only that component of flux in line with these coils will be available to produce an e.m.f.

The e.m.f.s so generated, will permit current to flow through the indicator or receiver coils to produce a resultant alternating magnetic field. Thus, the powered rotor of the receiver, in producing a magnetic field of its own, will react to this resultant field and line up with it. Hence, any movement of the transmitter rotor will be reproduced faithfully in the receiver.

A further look at the schematic arrangement in Fig. 10.22 will show no apparent difference between transmitter and receiver. In fact they are identical

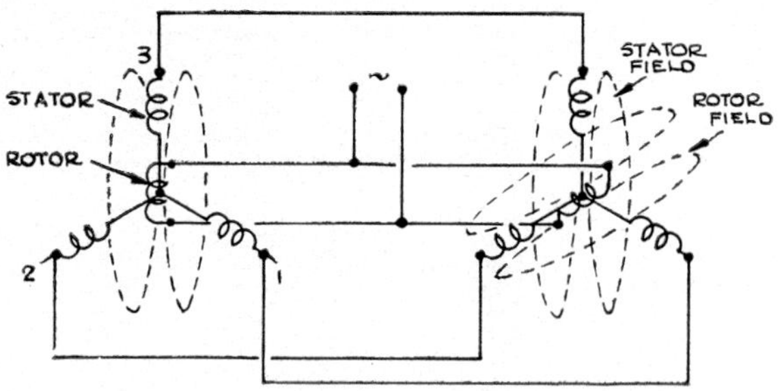

FIG. 10.22.

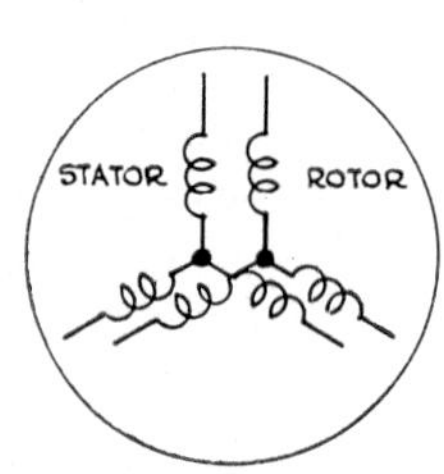

FIG. 10.23.

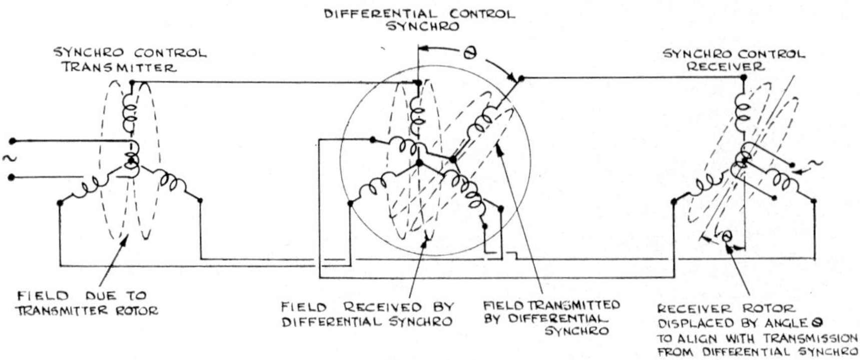

FIG. 10.24.

in principle, and therefore the receiver rotor will tend to drive the transmitter rotor into alignment. However, since the receiver rotor carries only a pointer, while the transmitting rotor is usually connected to a mechanical device, it will be evident that the resulting torque between rotors will move the one which is more lightly loaded; i.e. the receiver rotor. The transmitter and receiver are referred to as "synchros", the transmitter being a synchro torque transmitter and the indicator a synchro torque receiver.

Differential Synchros. In the normal synchro control system such as Aysynn it has been shown that the receiver follows the same angular displacement as the transmitter. In some instances however it is desirable that, while the receiver moves the same amount as the transmitter there should be a phase displacement between them. This is achieved by the differential control synchro. This is a device comprising a rotor and a stator each of which consists of a three-phase star-connected winding (Fig. 10.23). It is like a three-phase transformer where the stator can be regarded as the primary winding and the rotor as the secondary winding. When the rotor lies in the same plane as the stator, then, assuming no loss in the transformer action, the voltages at the primary will be repeated at the secondary. Should the rotor be moved through an angle in relation to the stator then the rotor voltage will be proportional to that angle.

Now consider the differential control synchro in action. It is connected into a synchro control system so that the stator of the differential synchro will be connected to the stator of the synchro transmitter. The rotor of the differential synchro is connected to the stator of the synchro receiver, so that a schematic of the circuit will be as shown in Fig. 10.24. Now assuming no angular displacement between rotor and stator in the differential synchro, the receiver rotor will be in line with the transmitting rotor. If the differential synchro rotor is now moved through a definite angle there will be a voltage

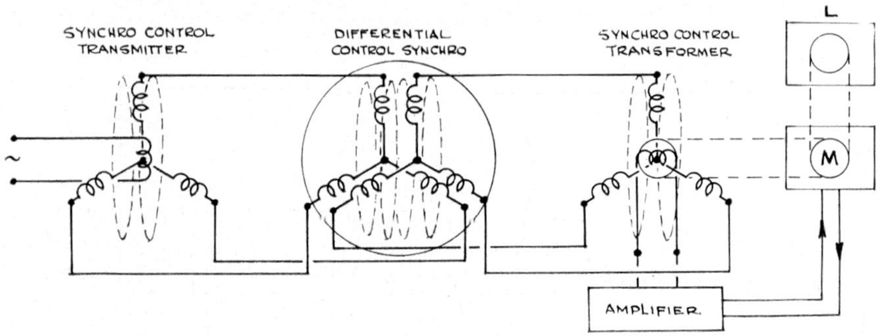

FIG. 10.25.

difference between stator and rotor. The change of voltage at the receiver will produce a torque to which the receiver rotor will respond. The system will now be such that the receiver rotor will respond to the transmitting rotor movement but there will always be an angular displacement between them equal to the angle between the rotor and stator of the differential control synchro.

Synchro Control Transformer. In some instances it is required that the receiver be used as a signal generator in order to produce a greater torque output than can be generated by the synchros themselves. The synchro control transformer consists of a three-phase stator and a two-pole rotor similar to those of the torque receiver. In this instance, however, the rotor is not supplied with power but will act as an inductive winding to the resultant alternating flux produced in the stator by the synchro control transmitter. The e.m.f. so generated can be used to operate an amplifier to drive a motor and so a load of some kind (Fig. 10.25). Here is a schematic circuit in which a synchro control transmitter is connected via a differential control synchro to a synchro control transformer.

If the transmitter (Fig. 10.25) rotor moves, so the resultant field in the receiver will change and an e.m.f. will be produced in the receiver rotor. This e.m.f. actuates the amplifier, drives the motor M to move the load L and, at the same time, drives the rotor of the receiver. When the receiver rotor has moved through the same angle as the transmitting rotor the e.m.f. to the amplifier decreases to zero. Thus the motor moves the load by an amount proportional to the angular displacement of the transmitter rotor. It will be noted that the transmitter and receiver rotors are at right-angles to each other for zero signal conditions. Should the differential control synchro be operated it will introduce a phase shift between output and input similar to that described for the synchro torque system.

In an earlier paragraph it was stated that the Aysynn system was used in the radio magnetic indicator, the radial magnetic selector and so on.

Before describing these instruments it should be mentioned that they are navigational instruments which depend on radio aids such as A.D.F. (Automatic Direction Finding) and V.O.R. (Very-high-frequency Omni-range System). A brief description of radio aids is therefore necessary to understand the use of the instruments as navigational aids.

The A.D.F. is a medium-frequency automatic direction-finding system and, as used today, is a simple aid comprising a ground transmitter and an airborne receiver. The principle of the system can be illustrated by Fig. 10.26. This shows a transmitter at *T* which is radiating energy in all directions; i.e. it is a Non-directional Beacon (N.D.B.).

An aircraft flying a course as indicated has come within range of the beacon and the receiver on board is picking up a signal via its aerial. In order to make use of this received signal to provide a bearing for the aircraft, the simple aerial takes the form of a loop. This loop is mounted vertically in the aircraft and is capable of being rotated through 360 deg. As the loop rotates so the signal received from the transmitter will vary, but it will be zero when the plane of the loop is "normal" (i.e. at right-angles) to a line *T–Y* between the aircraft and the transmitter (Fig. 10.27).

Thus in Fig. 10.27, an aircraft at *A* has its aerial "normal" to the direction of *T–Y* and therefore the signal from it is zero. The aircraft at *B* has its aerial in line with *T–Y* and will be receiving maximum signal. Thus, by rotating the aerial until its signal decreases to zero, the navigator can obtain

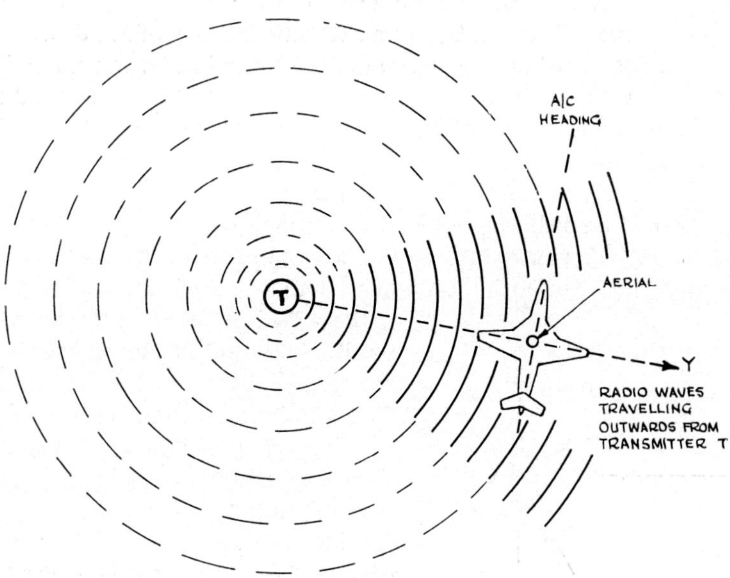

FIG. 10.26.

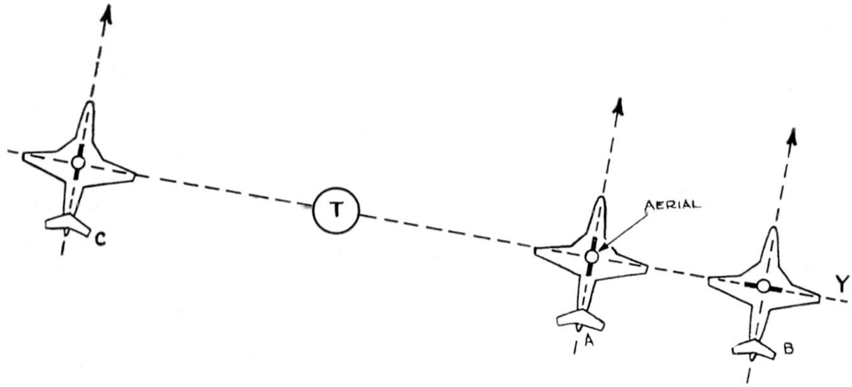

FIG. 10.27.

a bearing to the transmitter with respect to the aircraft head. An aircraft at C on the same heading as aircraft A also has its aerial normal to T–Y and therefore its signal will be zero. Thus there is a 180 deg. ambiguity in this simple system. In practice, the A.D.F. (by means of its aerial system) caters for this contingency and the ambiguity does not arise. A pointer fastened to the aerial can be made to move over a 360 deg. dial and thus show the angular displacement of the transmitter from the aircraft head. By arranging that the dial is rotated from a compass system the magnetic bearing of the transmitter is shown. Fig. 10.28 shows that the aircraft heading is 30 deg. and a bearing to the transmitter is 120 deg. left, or a magnetic heading of 270 deg.

V.O.R. is a very-high-frequency (V.H.F.) omni-range system which can also provide facilities of the same kind as A.D.F. although the method used is somewhat different. The V.O.R. system has an advantage over A.D.F. in that its transmission is V.H.F. and therefore subject to less interference than the medium-frequency range of A.D.F. Again, the system comprises a ground transmitter and an airborne receiver. The principle of the system is shown in Fig. 10.29. In (a) a transmitter T_1 with a non-directional beacon is radiating energy uniformly so that both aircraft 1 and aircraft 2 are receiving signals at the same instant. Since these signals are small alternating voltages they can be represented by sine waves as in (c). In fact, an aircraft within range of T_1, irrespective of position or heading, will receive exactly the same kind of signal as that of aircraft 1 and aircraft 2. Now consider the second aerial system as in Fig. 10.29(b). Suppose the energy transmitted from this aerial were "beamed" in

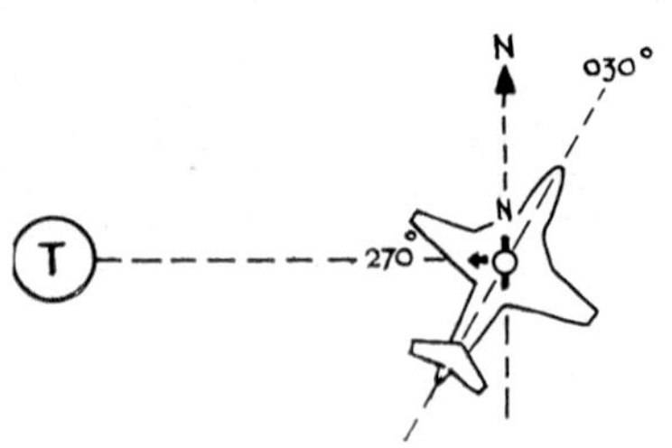

FIG. 10.28.

operated by a master compass to show aircraft heading against the lubber mark.

An additional feature on this instrument lies in the two switches at the top of the instrument. These are marked in the same manner as the pointers on the dial and enable the pointers to be switched to radio facilities in a number of ways: (*a*) both pointers to V.O.R. transmitters (as Fig. 10.37); (*b*) both pointers to A.D.F. transmitters; (*c*) No. 1 pointer to V.O.R. and No. 2 pointer to A.D.F.; (*d*) No. 1 pointer to A.D.F. and No. 2 pointer to V.O.R. A functional diagram of the R.M.S. showing the various synchros and operation of dials and pointers is shown in Fig. 10.38.

Finally, in order to show the operation of the R.M.S. the indications as shown in Fig. 10.37 are analysed and a schematic representation would be as in Fig. 10.39. This shows that No. 2 pointer of the R.M.S. is switched to V.O.R. and that, since the No. 2 pointer is vertical on the scale, the radio transmitter operating this pointer lies dead ahead of the aircraft on a heading of 325 deg. The aircraft can therefore be anywhere along the line *A–B*. However, No. 1 pointer is switched to V.O.R. and the radial of 055 deg. selected, as shown by the counter and by the crescent-shaped index. If the aircraft were also on a radial of 055 deg. to the second transmitter it would have to be at position P_1, whereupon No. 1 pointer would be at an angle of 90 deg. to No. 2 pointer.

The indication given in Fig. 10.37 is that No. 1 pointer, tuned to T_1, is at an angle of approximately 60 deg. and therefore the aircraft has to be at position P_2 to satisfy this presentation. Continuing on its present course will bring the aircraft to P_1, by which time No. 1 pointer will be opposite the crescent-shaped index. A turn to starboard of 90 deg. will cause the aircraft to turn on to a heading of 055 deg. and the compass card will have moved against the top index, carrying the crescent-shaped index with it. No. 1 pointer will now be upright, showing that the aircraft is flying the 055 deg. radial,

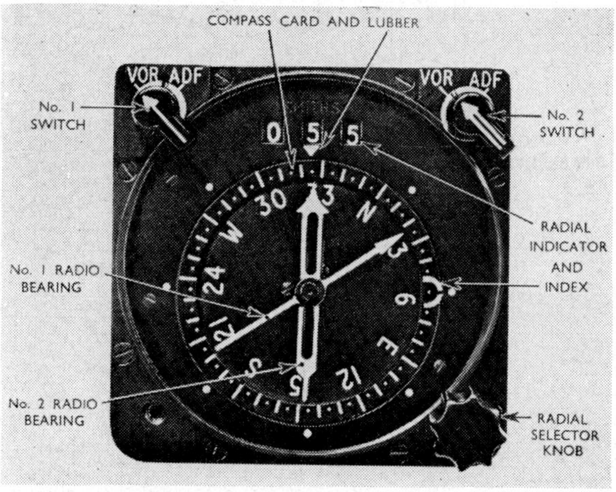

FIG. 10.37. *Radial magnetic selector.*

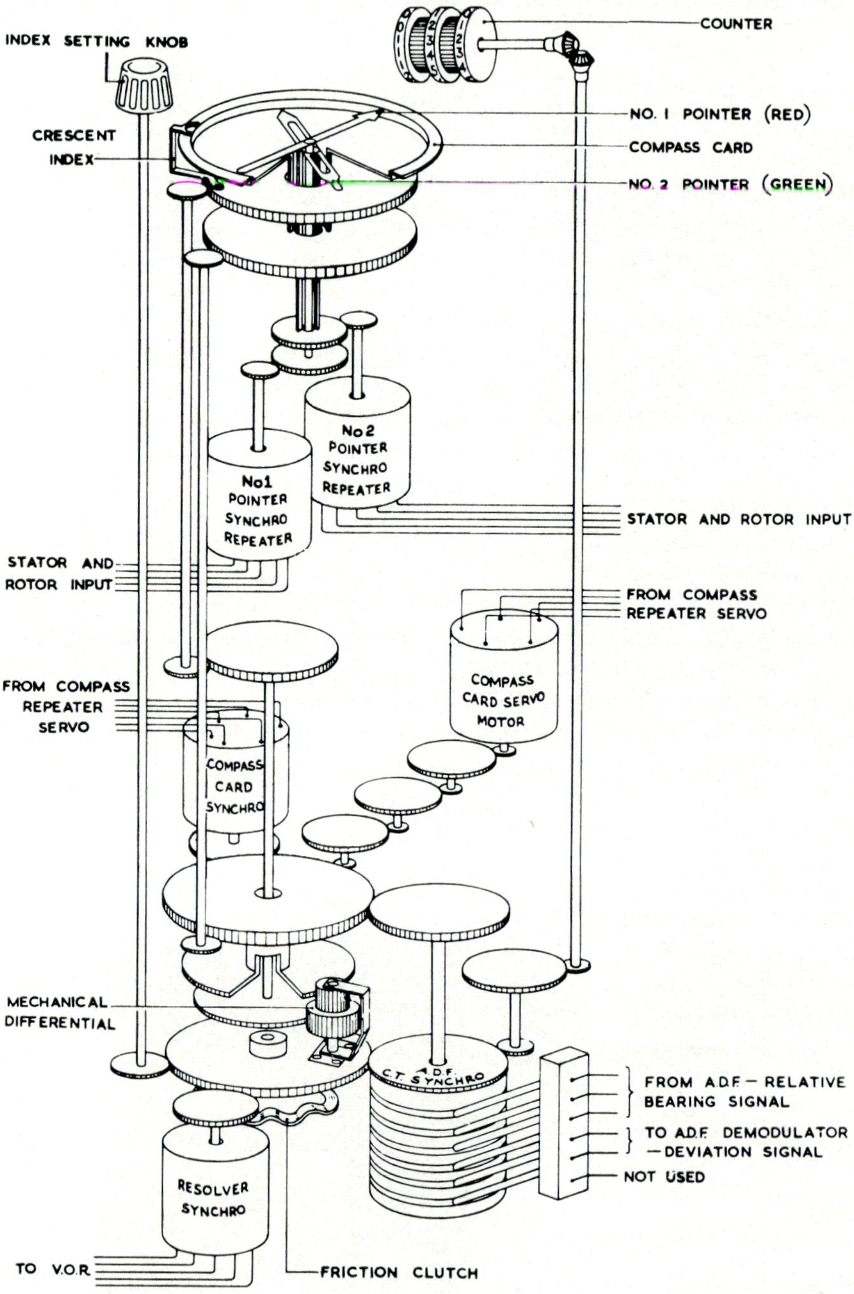
INDEX SETTING KNOB
CRESCENT
INDEX
COUNTER
NO. 1 POINTER (RED)
COMPASS CARD
NO. 2 POINTER (GREEN)
No 2
POINTER
SYNCHRO
REPEATER
No1
POINTER
SYNCHRO
REPEATER
STATOR AND ROTOR INPUT
STATOR AND
ROTOR INPUT
FROM COMPASS
REPEATER SERVO
COMPASS
CARD SERVO
MOTOR
FROM COMPASS
REPEATER
SERVO
COMPASS
CARD
SYNCHRO
MECHANICAL
DIFFERENTIAL
A.D.F.
C.T. SYNCHRO
FROM A.D.F. — RELATIVE
BEARING SIGNAL
TO A.D.F. DEMODULATOR
— DEVIATION SIGNAL
NOT USED
RESOLVER
SYNCHRO
TO V.O.R.
FRICTION CLUTCH

FIG. 10.38.

while No. 2 pointer will continue to move left to show an increasing angle between the two pointers.

Testing and Calibration. The testing of instruments in this category is mainly concerned with the testing of the various synchros employed. This is rather specialized work and no attempt will be made here to try to summarize the various procedures. Rather, one should refer to the manufacturer's literature when dealing with a synchro system.

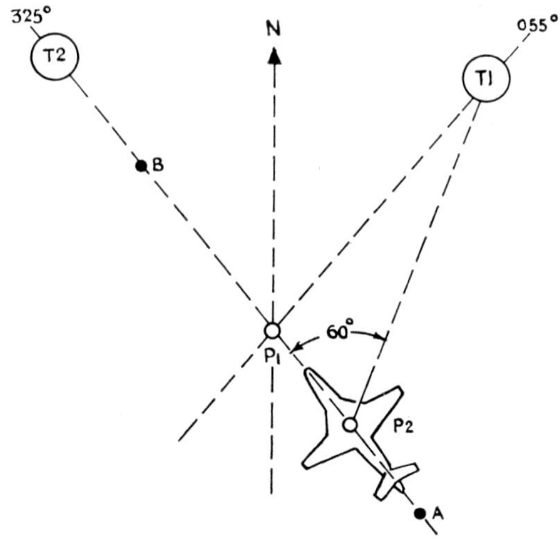

Fig. 10.39.

INTEGRATING FLOWMETER

The integrating flowmeter is another instrument which makes use of the Aysynn synchro system. It is used to measure the rate at which fuel is being consumed during flight. Additionally, an integrating device converts the flow-rate to give an indication of "fuel gone" and therefore, given the quantity or mass of fuel on board at the start of the flight, to enable the amount of fuel remaining to be calculated. The flowmeter consists of (*a*) a transmitter, which measures the rate of flow of the fuel, and (*b*) an indicator which records the rate of flow (see Fig. 10.40). The transmitter consists of an

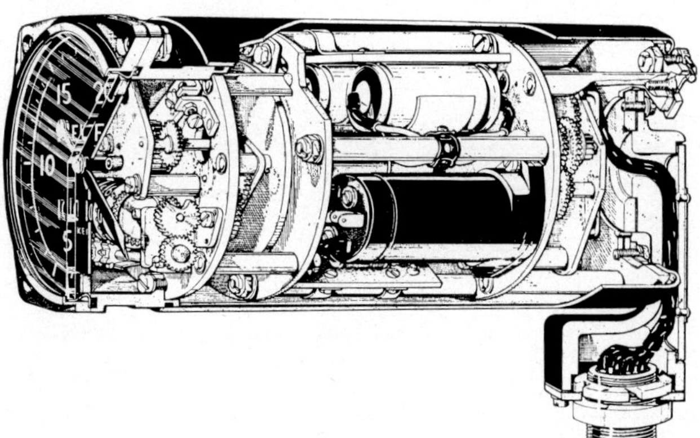

Fig. 10.40.

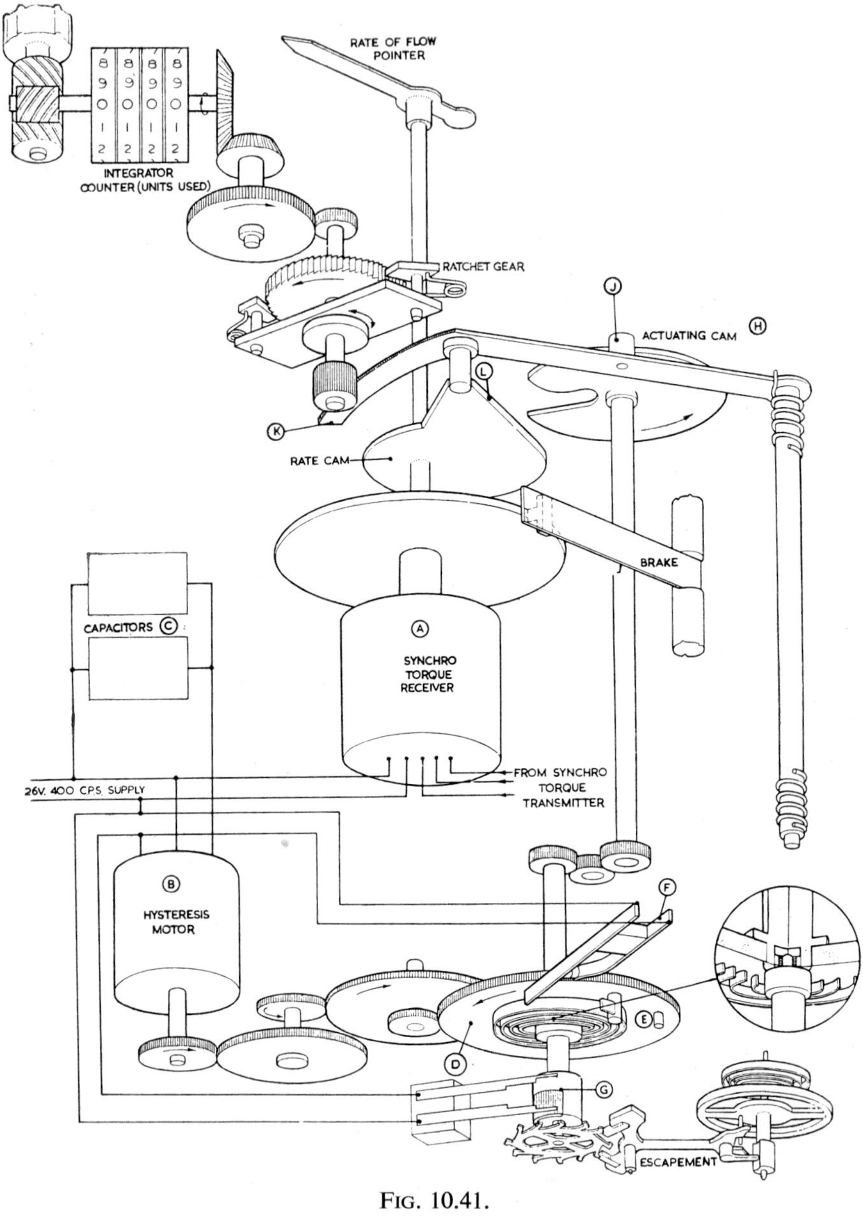

FIG. 10.41.

orifice through which the fuel passes at a rate dependent on the power output of the engine being supplied. Flow through the orifice creates a differential pressure across it, and this is used to displace a piston. The piston, in turn,

°C.	M/V	°C.	M/V
0	−0·80	300	11·41
20	0	350	13·49
50	+1·22	400	15·59
100	3·30	500	19·84
150	5·33	600	24·10
200	7·33	700	28·34
250	9·36	800	32·51

Suppressed-zero Thermometers. These instruments are sometimes used in place of the direct type of thermo-couple instrument for measuring jet temperatures. As the name suggests, the e.m.f. generated by the thermo-couples is suppressed so that the scale range is of the order 400–800 deg. C. thus giving greater sensitivity of reading. The suppression is achieved by feeding a voltage from the aircraft supply to "back-off" the thermo-couple voltage up to 403 deg. C. At this point thermal e.m.f. balances the suppression e.m.f.

The circuit is arranged as shown in Fig. 11.5. In this case the voltage compensator is connected to the aircraft supply and feeds out a constant supply of about 1·5 volts to the Wheatstone bridge network. Similarly, the four thermo-couples, connected in series, are feeding an e.m.f. proportional to temperature to the other two terminals of the bridge. It is arranged that at 403 deg. C. the opposing voltages are such as to balance so that the indicator reads this balance point of 403 deg. C. At temperatures lower than this value the thermo-couple e.m.f. does not balance the constant e.m.f. and the indicator pointer is therefore driven backwards off the scale. At temperatures greater than 403 deg. C. the thermal e.m.f. is such as to advance the pointer over the scale to indicate the thermo-couple e.m.f. and therefore the temperature.

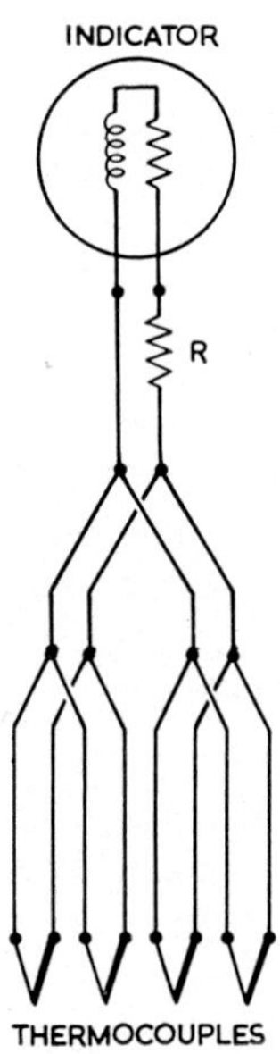

FIG. 11.4.

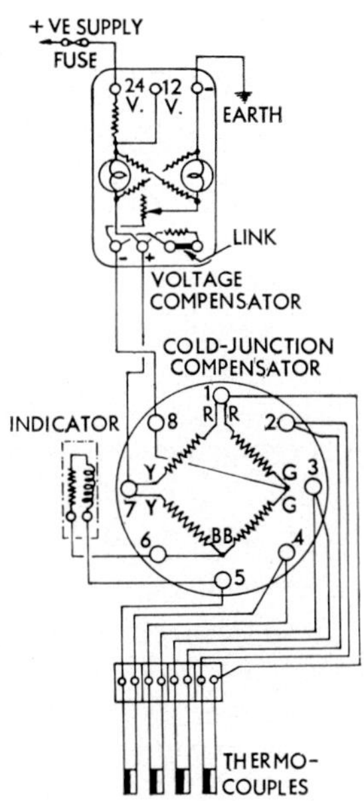

FIG. 11.5.

Thermo-couples. The chromel-alumel couples used in this and the previous system

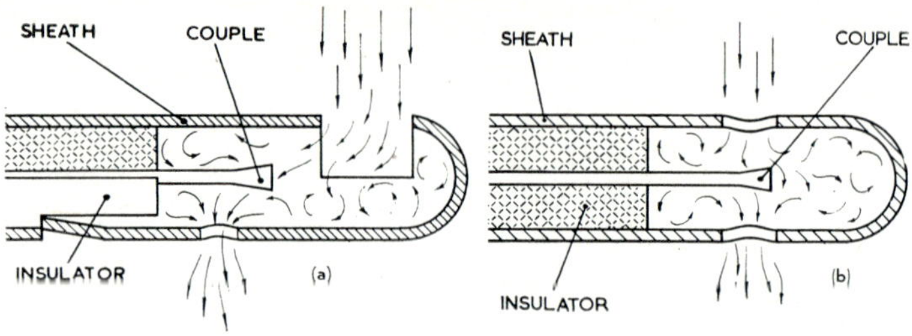

FIG. 11.6.

may be one of two kinds: (*a*) The *stagnation type* in which the leads are housed in a sheath which allows the exhaust gases to be arrested in the sheath; it is used where the exhaust gas velocities are high, such as in a pure jet engine (Fig. 11.6*a*). (*b*) The *rapid-response type* in which the leads are also housed in a sheath, but in this instance directly available to the gas flow. It is usually employed with the propeller turbine engine where gas velocities are lower than in the pure jet engine (see Fig. 11.6*b*).

Cold-junction Compensator. In the suppressed-zero type of instrument the Wheatstone bridge forms the cold junction of the system. It is necessary therefore that the compensation for changes of cold-junction temperature should be in this unit and not in the indicator, as previously. The compensation takes the form of a temperature-sensitive resistor in one leg of the bridge. Its effect is as follows. An increase of temperature at the cold junction will reduce the difference in temperature between hot and cold junctions thereby tending to reduce the indicator reading. However, this same increase of temperature will reduce the voltage from the supply across the bridge so that opposition to the thermal e.m.f. is reduced. Hence the system is independent of temperature changes of the cold junction.

Voltage Compensator. This can be connected to either a 12- or 24-volt source according to the aircraft supply. The appropriate supply is connected to a bridge circuit comprising two resistors and two lamps (which are non-linear resistors). The output voltage of approximately 1·5 volts is taken from this bridge network to the cold junction compensator. Should the input voltage to the voltage compensator vary, the change in the bridge network is sufficient to maintain the output of 1·5 volts.

Indicator. This is a normal millivoltmeter calibrated in Celsius degrees and having a range of 400–800 deg. C. It is connected to the cold-junction compensator by means of normal electrical leads. The cold-junction compensator compensates for the "extra" thermo-couple created when the indicator leads come into contact with the chromel-alumel leads at the junction. No compensation is necessary in the indicator itself. It is, therefore, not affected by ambient-temperature changes and the instrument does not need any setting up other than to see that it reads the neutral point of 403 deg. C. when disconnected from the circuit.

The indicator is tested as a millivoltmeter using the following table which shows temperature equivalents of millivolt inputs.

°*C.*	*M/V*	°*C.*	*M/V*
400	−0·48	600	33·68
403	0·00	650	42·17
450	+8·09	700	50·64
500	16·64	750	59·09
550	25·17	800	67·52

The tolerance is of the order of 1 to 2 per cent. of full scale according to the reading of the particular instrument; e.g. below 600 deg. C. the tolerance is usually ±1 per cent.

GAS TURBINE TEMPERATURE CONTROL

The maximum output power from a gas turbine is obtained when the engine is running at a very high temperature. However, if one considers the effect of this high temperature on the turbine blades it will be realized that operating temperature must not be considered merely in terms of output power but also in terms of the limiting temperature that can safely be withstood by the turbine blades. The temperature of the engine can be controlled by controlling the fuel flow; thus it is possible for a pilot or flight engineer to read the exhaust-gas temperature indicator and manually control the fuel flow to control temperature. Obviously, the drawback to such a procedure occurs from the time-lag in reading the instrument and applying the control, and from the constant need to monitor the exhaust-gas temperature indicator. Additionally, control should be applied even for small variations of temperature which may be difficult to observe on the standard indicator. Thus, a temperature controller is a desirable feature of the aircraft carrying gas-turbine engines.

The principle of the temperature-control system can be shown as in Fig. 11.7. A number of thermocouples, *A*, are inserted into the jet exhaust pipe so that an e.m.f. is generated

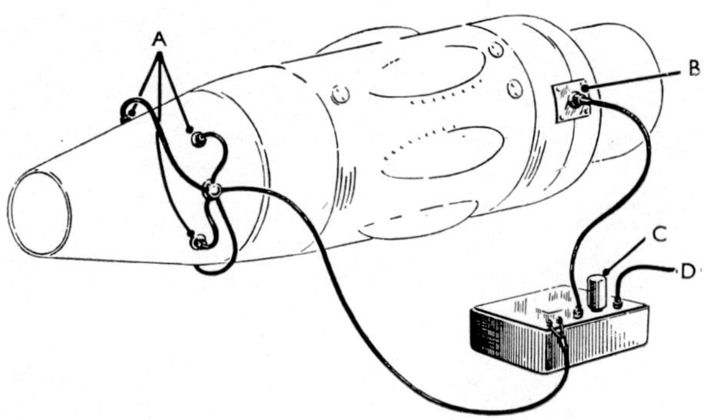

FIG. 11.7.

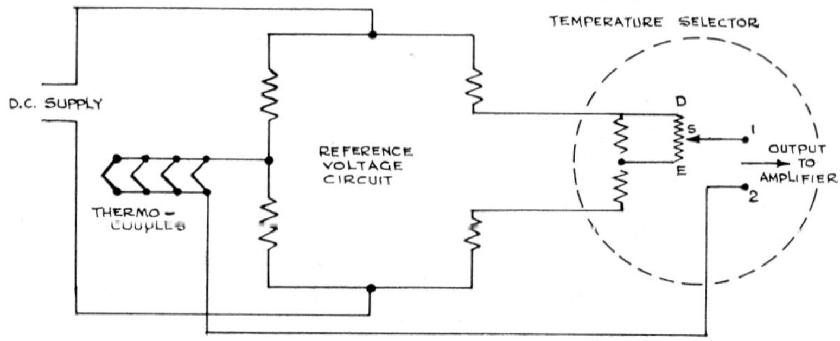

FIG. 11.8.

according to the temperature of the exhaust gases. This e.m.f. is applied to an amplifier operating a fuel control valve, *B*, which thereby determines operating temperature. A temperature selector, *C* is fitted to the amplifier to allow optimum operating temperature conditions to be selected for the particular turbine. The amplifier is fed from a supply *D* at 115 volts, 400 c.p.s.

Thermo-couples. These are of the chromel-alumel type described earlier. Usually four of these couples are arranged radially about the jet tail pipe and they are connected in parallel. A terminal block on the bulkhead takes all the thermo-couple leads and provides a means of connecting the couples to the amplifier. The terminal block thus becomes the cold junction of the system which requires to be corrected for changes in ambient temperature. This is done by means of a cold-junction compensator winding.

Temperature Selector. The temperature selector consists of a resistance network which can be adjusted to give the required output from a reference-voltage circuit. This output is opposed by the thermo-couple e.m.f. and therefore the setting of the selector determines the operating temperature

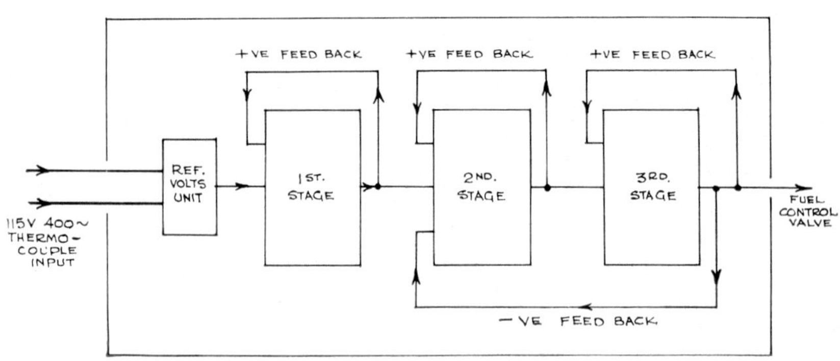

BLOCK DIAGRAM OF AMPLIFIER

FIG. 11.9.

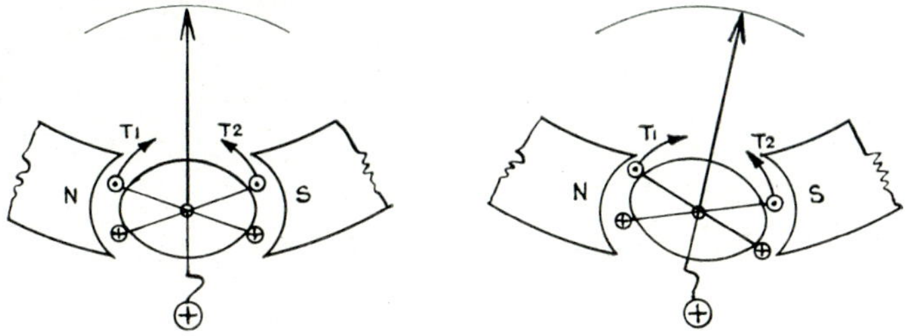

FIG. 11.15.

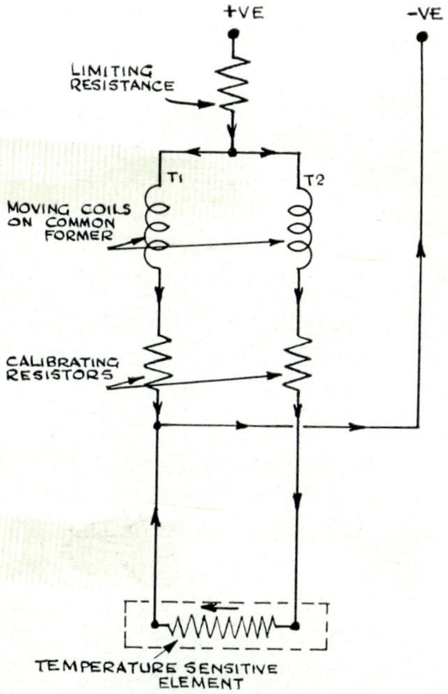

FIG. 11.16.

the coil carrying the stronger current is moving into a weaker part of the field. Thus, a new balance is reached between T_1 and T_2 and the angular deflection of the pointer is proportional to the change in temperature.

Change in Supply Voltage. Should the supply voltage change, there will be a reduction of current in each coil but the *ratio* of currents remains the same and therefore the instrument is independent of voltage. This type of instrument is known as a "ratio-meter". (Since the coil is fitted with a weak spring to pull the pointer off the scale in the event of a supply-voltage failure the instrument is only independent of supply voltage so long as any changes are small. This is due to the fact that the movement balances at a point where T_1 is equal to T_2 plus hairspring torque. It will be seen therefore that as T_1 and T_2 decrease with decreasing voltage, the hairspring torque has greater effect. However, within the limits of the usual change, say 26 to 29 volts, the instrument can be regarded as accurate.)

The instrument is tested by connecting it to a decade resistance box. The two terminals of the decade box are connected to the two indicator terminals which normally carry the resistance-element leads. Resistance values of the box are set to the figures given in the table on page 157 and the indicator should show a temperature equivalent to the resistance within a tolerance of 1 to 2 per cent according to the scale range.

CHAPTER 12

Electrical Pressure Gauges

IN the ratiometer type of thermometer described in Chapter 11 it will have been noted that the indicator measures a change of resistance. It is possible to use the ratiometer for other forms of measurement by making use of appropriate resistance elements, as described in this chapter.

RATIOMETER PRESSURE GAUGE

This is a d.c. electrical pressure gauge comprising a pressure-sensitive resistance element and a ratiometer-type indicator which is independent of voltage variations. In appearance the pressure-sensitive resistance element is much like the Desynn pressure transmitter as it comprises a bellows pressure unit and a resistance transmitter similar to the "micro" transmitter. However, the electrical circuit of the transmitter is simpler than that of the Desynn and usually consists of two resistance bobbins side by side and connected in parallel. A change of pressure in the bellows unit causes two brushes connected electrically to move over the resistance bobbins. Thus, the resistance change is proportional to the pressure change.

In some cases a warning contact is included in the transmitter so that at low pressures the indicator shows the value of the pressure and also causes a warning lamp to light to draw attention to the indication. When the warning contact is used, the transmitter houses only one resistance bobbin; the other is replaced by a metal contact. The circuits of the ratiometer pressure gauge, with and without warning contacts, can be shown schematically (Fig. 12.1).

The indicator is identical in principle and operation with the ratiometer movement used for temperature measurement. It is not usual, however, to find this indicator mounted in its own instrument case. Rather, is it made in the form of a dual instrument where the two separate movements can be used to show two different pressures, or with one side showing pressure and the other side showing temperature. Thus it is possible to indicate the pressure and temperature of the lubricating oil for one engine in one instrument case. It will be noted from Fig. 12.1 that the circuit for pressure indicators is a little different from that of the temperature indicator in that the current in both coils changes with changes of pressure.

Testing. In order to test the pressure indicator it is necessary to have two decade resistance boxes, the circuit being as in Fig. 12.2. The values set on the two resistance boxes will vary with the range of the indicator under test.

There are two ways of testing the transmitter. In the first, the transmitter is connected to an indicator, the accuracy of which has been verified by the decade-resistance test. Now apply progressive pressures to the transmitter by

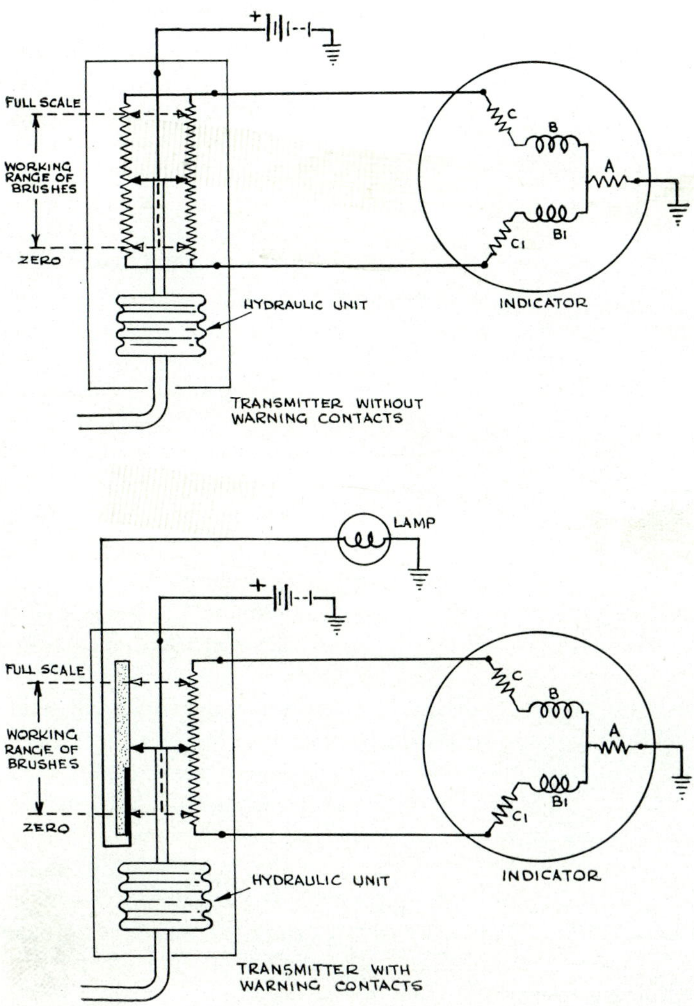

FIG. 12.1.

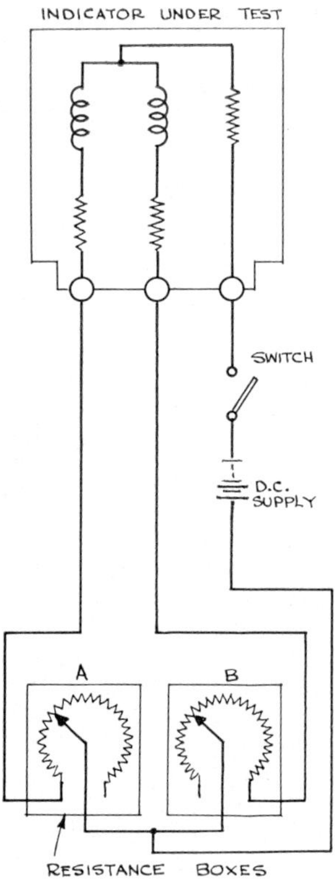

FIG. 12.2.

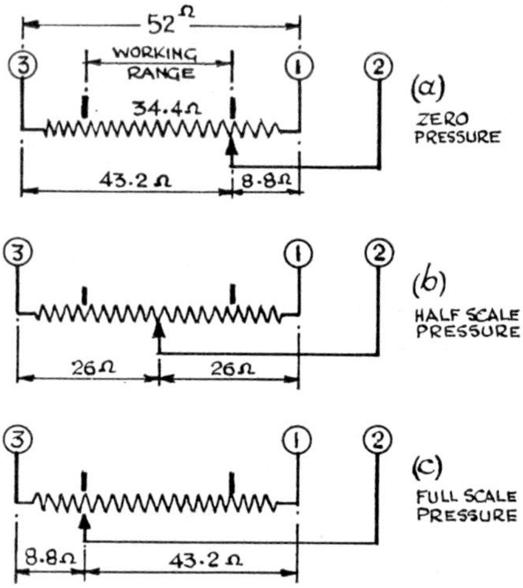

FIG. 12.3.

means of a dead-weight tester and note the indicator readings. Knowing the accuracy of the instrument will enable the transmitter accuracy to be determined.

A more accurate method, but one requiring more test equipment, is as follows. Measure the value of the resistance of the transmitter between terminals 1 and 3 (see Fig. 12.3). The value should be of the order of 50 ohms. The working range of the wiper over the resistor is 33·4 ohms. Supposing, therefore, the total value of resistance between the terminals was found to be 52 ohms, then there is a resistance of (52 — 34·4)/2 ohms; i.e. 8·8 ohms at each end of the working range.

Now, with zero pressure on the transmitter, the wiper will be at one end of the working range and therefore the resistance between the wiper and lower end of the resistor (terminal No. 1) should be 8·8 ohms. Similarly, the resistance between the wiper and the other end of the resistor should be 52 — 8·8 = 43·2 ohms. If pressure is applied to the transmitter equal to half scale range then the value of the resistance between the wiper and either end of the resistor should be 26 ohms. Apply pressure now equal to full-scale range and the resistance values obtained for zero pressure will again be obtained, but in the reverse order. The schematic diagram in Fig. 12.3 illustrates this. The indicator tolerances are ±2 per cent of scale range, while transmitter tolerances are ±0·4 ohm at zero and ±0·8 ohm elsewhere over the range.

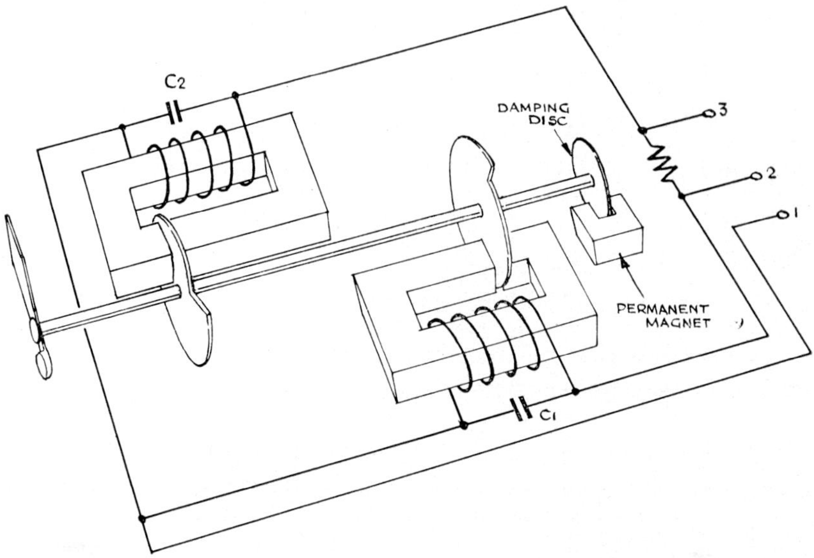

FIG. 12.6.

flux and this flux reacting with the flux across the air-gap, produces a torque on the vane, which therefore starts to move about its pivoting point.

If it is realized that the same thing is happening to the other vane in the second air-gap and that the torque on this vane is opposing that of the first vane, then the direction of rotation will be seen to depend on the ratio of currents in the two coils. In order to achieve a balance of torques, the vanes are shaped like cams and the rotation is such that the radius of the vane increases in the air gap with the weaker current and increases with the stronger current.

Compensation. Since the position of the vanes, and therefore the pointer, is a measure of the ratio of current the system is independent of voltage variations within the usual aircraft tolerances. An increase of frequency would produce an increase of reactance of the indicator coils. The reactance of the capacitors in parallel with the coils, however, decreases with increasing frequency so that the impedance of the network remains substantially the same. The effect of temperature on the system is to change the value of the impedances in the network. Since these impedances are not equal (except at half-scale) the effect of temperature is to change the ratio of impedances. The change is such that an increase of temperature would reduce the ratios to reduce the scale range. It is necessary therefore to provide compensation which will tend to increase the scale range as temperature increases. A temperature-sensitive resistor is therefore connected across the indicator

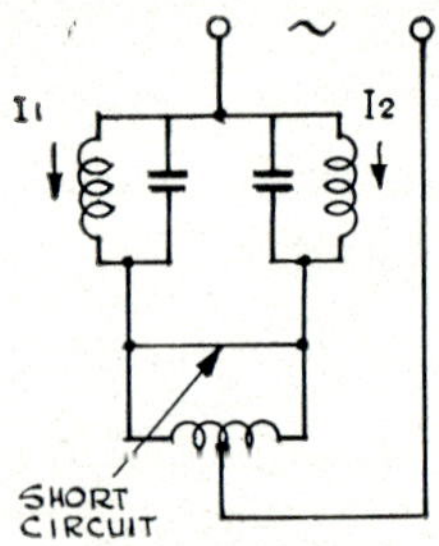

FIG. 12.7.

coils. The principle of operation of this compensator is as follows.

If a straight conductor is placed across the indicator coils then these coils are short-circuited (Fig. 12.7), and, irrespective of the transmitter output, the indicator would rest at a half-scale position since the currents in the coils are equal. Because the pointer is fixed in this way, it may be said that the range of the instrument is zero; i.e. its minimum and maximum points are identical (mid-scale). Now if the short-circuiting wire is removed, the pointer will move so that its minimum and maximum points are removed from each other; therefore the range of the instrument has increased.

Now a circuit with a high resistance is similar to one with an open circuit. Therefore, if a high resistance is placed across the two coils the range of the instrument is very nearly that of the same indicator on an open circuit. An increase of temperature—which reduces the ratio of impedances—reduces the scale range. The same increase of temperature, however, increases the sensitive resistance across the coils, which accordingly tends to increase the range. The two reactions are opposite, and the effects of temperature are thereby minimized.

Testing the Indicator. The a.c. ratiometer is tested by connecting to it a test set comprising standard inductors (see Fig. 12.8). The normal supply voltage is connected to the system and the indicator readings are compared with the values for the particular inductors used. The inductors required are four in number and their values are given in Table 12.1. It is possible to obtain a test set with these standard inductors appropriately wired with suitable switches and plugs to ease the process of testing. The test set is obtainable from the instrument manufacturer and is shown in Fig. 12.9.

Transmitter. The simplest way to check the transmitter is to wire it to an

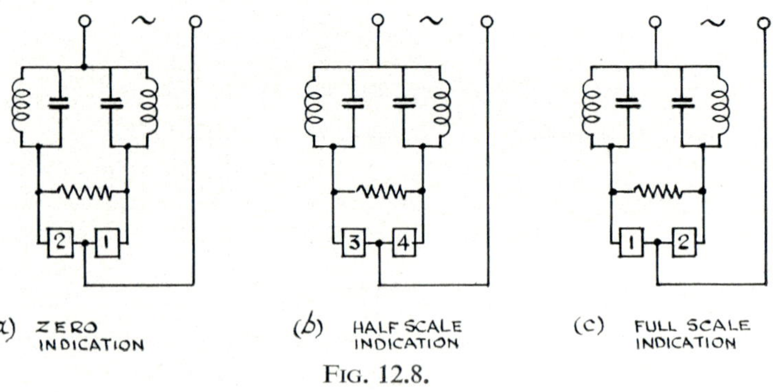

FIG. 12.8.

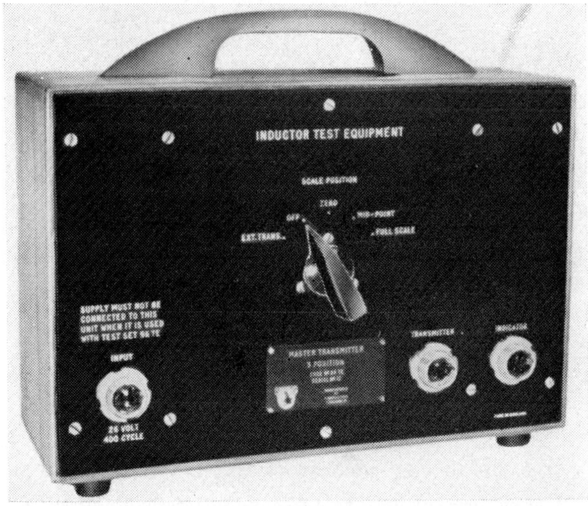

FIG. 12.9. *Indicator test equipment master transmitter. Sockets are provided for the input and for connecting the transmitter and indicator.*

indicator whose accuracy has been verified by the test set. Applying pressure to the transmitter by the usual dead-weight tester will provide indicator readings from which the transmitter errors can be deduced. The permitted tolerance for transmitters and indicators is ±2 per cent.

Table 12.1

INDUCTOR VALUES

Inductor	*Inductance (mH)*	*A.C. Resistance (ohms at 20 deg. C.)*
1	53·5 ± 0·03	56·731 ± 0·3
2	26·5 ± 0·03	50·651 ± 0·3
3	40·0 ± 0·03	53·748 ± 0·3
4	40·0 ± 0·03	53·748 ± 0·3

Portable Test Set. A portable test set is available from the manufacturer for testing transmitters and indicators. This portable device, however, is merely for checking the functioning of the system and in no way verifies its accuracy, as does the test set shown in Fig. 12.9.

CHAPTER 13

Electrical Measurement of Engine Speed

THE electrical method of measuring engine speed may be by means of either a d.c. or an a.c. system. In either case the system comprises an engine-driven generator, and an indicator which shows the state of the generator at some particular engine speed.

The d.c. system is hardly seen on modern aircraft and it is sufficient to say that it comprises a d.c. generator whose output voltage is proportional to engine speed. The indicator consists of a moving-coil voltmeter which records the voltage output of the generator. The indicator can therefore be directly calibrated in r.p.m. The main disadvantage of this system lies in the fact that it is voltage-sensitive, so that a change in the permanent magnets employed in generator and indicator would materially affect the readings obtained.

In the a.c. system the generator may be driven by a short length of flexible drive or, better still, by being flange-mounted to the engine. The generator is simple in operation and consists of a stator and a rotor. The stator is usually three-phase, star-wound. The rotor is a permanent magnet, which may be of the two-pole or four-pole type. It is driven by the engine at the same speed as the crankshaft in a piston engine, but at a sub-multiple of the speed of the turbine in a jet engine. The reason for this is to limit the mechanical loads on the generator by fixing a maximum speed for the rotor of 5,000 r.p.m. in the case of the two-pole type, which is the more convenient to study. Therefore, in the case of a turbine whose maximum speed does not exceed 10,000 r.p.m. the generator speed is reduced by a half, while in the case of a maximum turbine speed of 20,000 r.p.m. the generator speed is reduced to a quarter of the engine speed.

Where a flexible drive is employed, usually on the piston engine, it too may run at a sub-multiple of crankshaft speed, but a gear-box in the generator will restore the rotor to the same speed as the crankshaft. The reduction of flexible-drive speed increases its life and efficiency. The rotor may run in plain sintered-bronze bearings or in ball races and when flange-mounted to the engine will have an oil seal fitted (see Fig. 13.1).

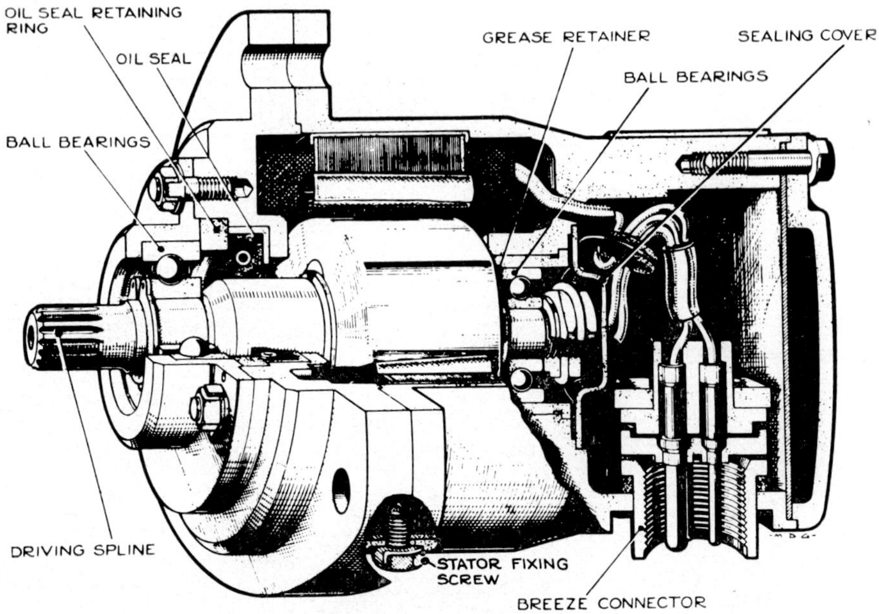

FIG. 13.1. *Flange-mounted generator for engine-speed indicator.*

Principle of Operation of Generator. Consider the simple generator consisting of a rotating permanent magnet and a single coil of wire (Fig. 13.2). With a magnet in position (*a*) the flux passing through the coil induces an e.m.f., with one end of the coil at a higher potential than the other. At (*b*) after 90 deg. of rotation, no flux is passing through the coil and the e.m.f. decreases to zero. At (*c*) a further rotation of 90 deg. (i.e. 180 deg. in all) sees the flux passing through the coil once more but in the reverse direction to that at (*a*) so that the e.m.f. is reversed. In (*d*) after 270 deg. rotation, the e.m.f. once more decreases to zero, after which the cycle is repeated. Thus, the magnet induces an e.m.f. the value of which depends on the strength of the magnet and the number of turns on the coil. The number of alternations per second (or cycles per second) is determined by the number of times the magnet rotates per second. Therefore, the frequency is proportional to magnet speed, and if the magnet is driven by the crankshaft or turbine, then the frequency of the e.m.f. must be a measure of engine speed.

A generator with a three-phase star-wound stator can be shown schematically as in Fig. 13.3. This shows that the magnet is producing an e.m.f. in each of the three coils, the frequency of the e.m.f. in each case being dependent on the magnet speed. The peak value of the e.m.f. in each coil will occur 120 deg. after the preceding coil since these are also displaced by

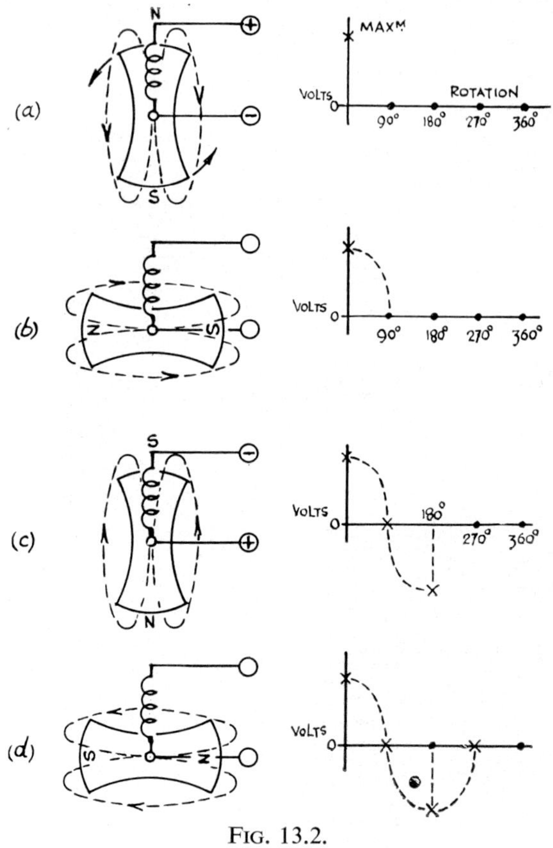

FIG. 13.2.

120 deg. In practice it is possible to connect the ends of the coils 1*A*, 2*A* and 3*A* to form a common star point, and the generator output leads will be shown as 1, 2 and 3.

Indicator. The indicator (Fig. 13.4) consists of a three-phase star-wound stator, a rotor and, attached to the rotor, a permanent magnet with drag-cup assembly. The rotor is driven by a magnetic field set up in the stator by the generator e.m.f., the rotor speed being proportional to frequency and therefore to engine speed. A permanent magnet mounted on the rotor produces eddy currents inside the drag-cup assembly, the field produced by the eddy currents providing a torque to rotate the cup against a control spring. The angular movement of the cup is proportional to engine speed and, by means of gears, one or more pointers may be made to move over a dial calibrated in r.p.m.

Principle of Operation of the Indicator. The indicator stator is similar to

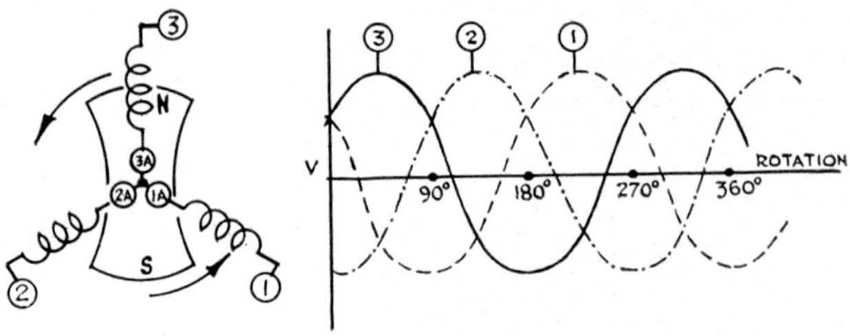

FIG. 13.3.

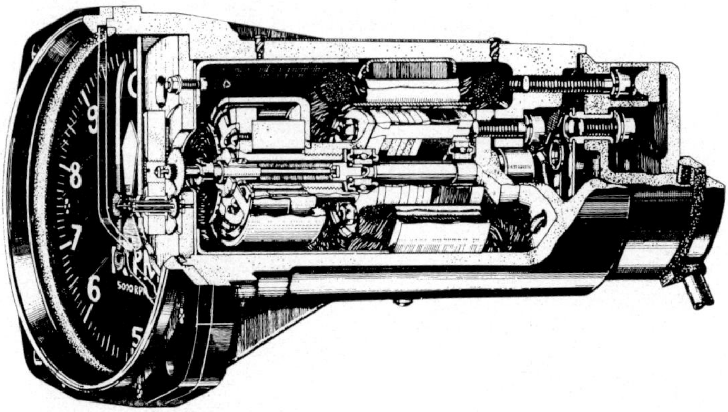

FIG. 13.4. *Engine-speed indicator with details of rotor, stator and drag-cup assembly.*

that of the generator and is connected to it by the three-phase leads. The generator e.m.f.s are therefore supplied to the three indicator leads and produce currents in the stator coils, the magnitude and direction of the currents being dependent on the value of the e.m.f.s (see Fig. 13.5). At the time t_1 the value of the e.m.f. in phase 3 is maximum and positive, while the e.m.f.s of phase 2 and phase 1 are negative. Therefore, current flow in the direction indicated produces a resultant magnetic field R_1.

At the second instant in time t_2 60 deg. later, it will be noted that: the e.m.f. in phase 3 has decreased in value; the e.m.f. in phase 2 has reversed in direction and is equal in value to the e.m.f. in phase 3; and that the e.m.f. of phase 1 has increased to maximum negative value. The distribution of currents in the stator will now be as shown in Fig. 13.6. These currents produce a resultant magnetic field R_2 which will be found to be displaced from R_1 by 60 deg.; i.e. the same angle through which the rotor of the generator has moved. If this process were continued it would be found that the resultant field could be imagined as one which is rotating and, more important, is

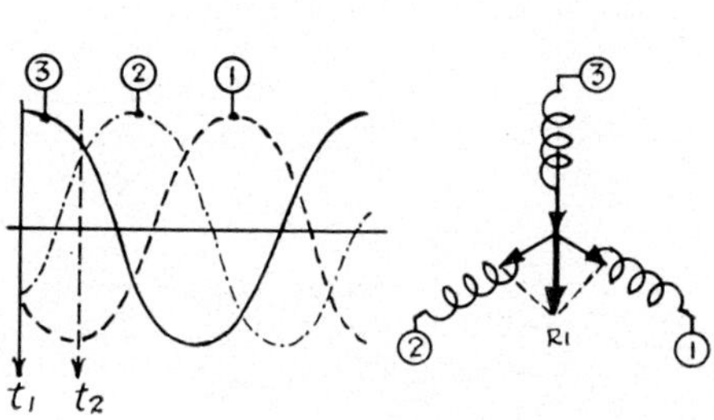

FIG. 13.5.

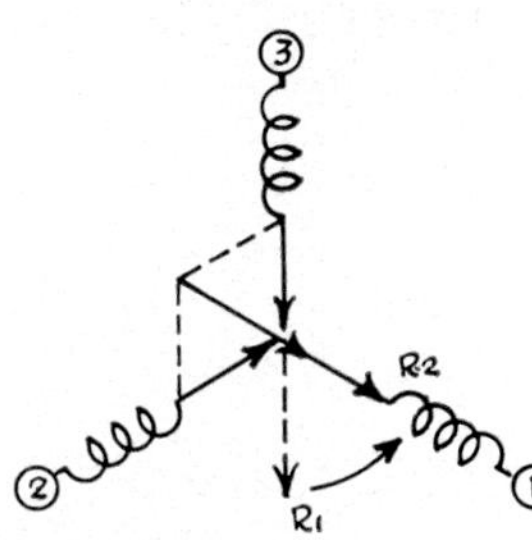

FIG. 13.6.

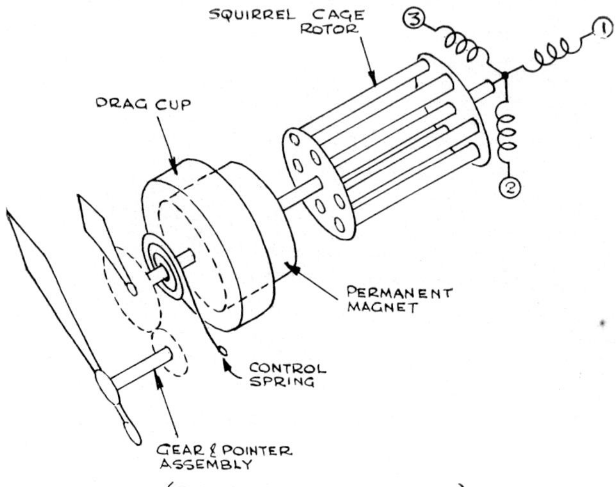

FIG. 13.7.

rotating at a speed dependent on the frequency of the generator and therefore on engine speed.

It now becomes necessary to convert this rotating magnetic field into something more tangible. This is done by housing a squirrel-cage rotor (Fig. 13.7) within the indicator stator. This rotor consists of two circular end plates which are joined together by a number of copper rods. The rotating field produces currents in these rods which, in turn, causes a magnetic field to be set up around them. The reaction of this field with the rotating field produces a torque on the rotor which makes it revolve in the direction of the rotating field at a speed dependent on the frequency of the generated e.m.f.s. A permanent magnet carried by the rotor moves in a drag-cup assembly and because of the eddy-currents, causes the cup to rotate in the direction of the rotor.

Testing. Generators are tested by running them at a given speed and

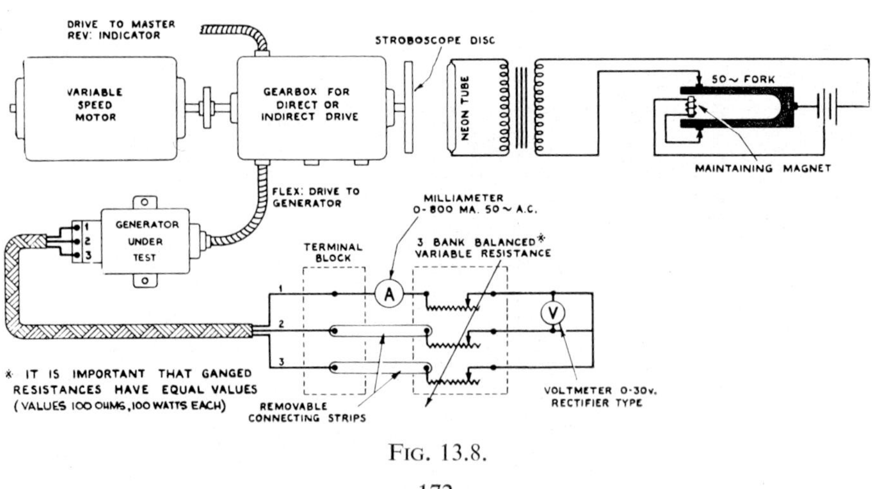

FIG. 13.8.

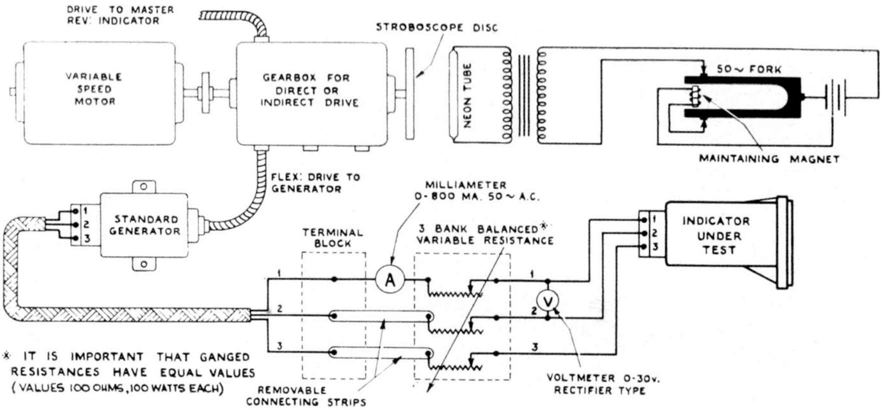

FIG. 13.9.

checking the voltage output on a given load. The circuit arrangement can be shown schematically as in Fig. 13.8. The generator is run at a speed of 3,000 r.p.m. and the ganged resistors are moved until a line current of 0·5 ampere is shown on the ammeter. The voltmeter reading between any two lines should now be 19½–21 volts. It should be noted that although this test is carried out, the figures given are not critical where only one indicator is used with a given generator (as is usually the case). It is sometimes necessary, however, to connect two indicators in parallel across one generator, and the power output should then be in accordance with the figures just quoted.

Indicators are tested for accuracy (Fig. 13.9) by connecting them to a generator of known characteristic, the generator then being run at known speeds. The indicator, of course, should agree with these known speeds, which are derived from a stroboscope. In order to test for rotor freedom in the indicator a further test is necessary. This involves the connection of a three-gang variable resistor in the circuit between indicator and generator. An ammeter is connected in one line and a voltmeter across any two lines. The generator is run at 3,000 r.p.m. with the ganged resistance at its maximum value, thereby reducing the input to the indicator. The indicator accuracy suffers, and the pointer will probably be oscillating on the dial. The resistance value is now reduced until it is noted that the instrument pointer moves forward to give a steady reading, thus indicating the fact that the rotor has synchronized with the generator. At this point the voltmeter reading should not be more than 15 volts, the ammeter reading being not more than 0·25 ampere. (In some instances the manufacturer may suggest a direct connection between indicator and generator and quote a maximum reading at which the indicator shall synchronize with the generator.)

Adjustments. As a general rule there are no adjustments on the generator,

FIG. 13.10. *Bryans bench-type tachometer tester.*

except that the permanent-magnet rotor is demagnetized to a value which ensures the test figures given earlier; i.e. 0·5 ampere, and 19½–21 volts.

If the readings of the indicator do not agree with the values given by the test equipment it becomes necessary to adjust the calibration. This consists of varying the value of eddy currents in the drag-cup assembly for a given rotor speed. The method by which this is done varies according to the manufacturer. In the Kelvin-Hughes indicator, for example, the magnet strength in the indicator remains at a fixed value and the eddy currents are varied by moving the drag-cup assembly relative to the magnet. In the case of the Smith type of instrument the reverse is true; i.e. the relative position of drag-cup assembly and magnet remains the same, but the magnet strength is varied by demagnetization (in the case of overreading).

Test Equipment. It will be noted that the equipment used for generators and indicators is the same in each case. Provision is made for checking one or the other by means of suitable switches and so on. The whole of the stroboscopic equipment can be quite compact and may occupy little space on the work-bench. A typical set is shown in Fig. 13.10.

STROBOSCOPE

The principle of the stroboscope was outlined in Chapter 8, but details of practical instrument will now be given. The earlier discussion suggested the idea of a ball rotating on the end of an arm (Fig. 8.5). Suppose that a device of this kind were housed in a box which included a small window, the interior of the box being normally quite dark so that it was not possible to see the rotating ball. Imagine now that a lamp is provided to illuminate the ball, the lamp being controlled by a switch enabling the illumination to be switched on and off at any instant. If the light were flashed on and then off at the same instant once in every revolution of the ball, then the ball would always be seen in the same position. Hence, if the speed at which the light flashes is known, then the revolutions per minute of the ball will be known. The flashing light provides the datum against which the speed is measured.

Neon Tube. A filament lamp is not very suitable for use in a stroboscope owing to the relatively slow heating and cooling of the filament. A neon tube is therefore used and this is made to flash by means of a "tuning fork" (see Fig. 13.9). This is a device which vibrates at a frequency determined by its structure. Its period of vibration is maintained by allowing it to make and break a d.c. circuit which provides an electro-magnet with sufficient energy to attract the prongs of the fork when the switch is closed, and is de-energized when the switch is opened. The fork will thus vibrate at its natural frequency so long as the d.c. supply is available.

As the fork vibrates, so a second pair of contacts interrupts a supply to a transformer, thereby creating a high potential across the secondary winding, which causes the neon tube to light. The neon tube therefore lights at the same speed as the fork vibrates, and this is usually 50 c.p.s. (i.e. 3,000 times per minute). Now if the rotating ball were running at 3,000 r.p.m., then it would be illuminated by the neon flash when it occupied the same position during each revolution and, to the observer, would appear stationary.

Suppose the speed of the ball were reduced to 1,500 r.p.m., while the light still flashed at 3,000 times per minute. The ball would now be illuminated twice every revolution and the observer would apparently see two balls at diametrically opposite points. Similarly, if the ball rotated at 750 r.p.m. it would be illuminated every quarter-revolution and would therefore provide a picture of four balls each displaced from the other by a quarter-revolution. It is possible therefore to determine the speed of the ball from the "picture". Supposing the picture shows six balls equally spaced; this means that the rotating ball is being illuminated six times every revolution and its speed is therefore $3{,}000/6 = 500$ r.p.m. Other values can be worked out accordingly.

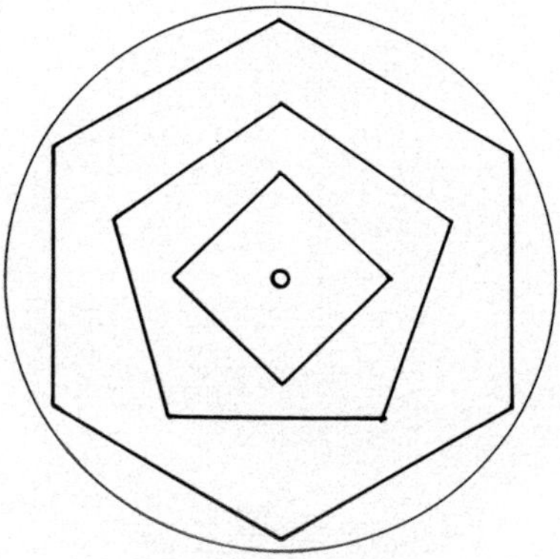

FIG. 13.11.

Stroboscopic Disc. The more practical way of interpreting the picture is to use a disc such as is shown in Fig. 13.11. On this is drawn a four-sided, a five-sided and a six-sided figure. With a neon tube flashing at 3,000 times a minute on this rotating disc, the square will appear stationary at 750, 1,500, 2,250 and 3,000 r.p.m.; the pentagon at 600, 1,200, 1,800, 2,400 and 3,000 r.p.m.; and the hexagon at 500, 1,000, 1,500, 2,000, 2,500 and 3,000. These figures will provide a means of determining the various speeds of the disc:

Square	*Pentagon*	*Hexagon*
3,000	3,000	3,000
2,250	2,400	2,500
1,500	1,800	2,000
750	1,200	1,500
	600	1,000
		500

Thus, it would be seen that all the figures appear stationary at 3,000 r.p.m and the square and the hexagon will both appear to be stationary at 1,500 r.p.m. By introducing other types of figures it is possible to determine variations in speed of the disc of 100 r.p.m.

Referring again to Fig. 13.9, the disc is being driven by a motor at the same speed as a standard generator. The indicator speeds should therefore conform to the speeds read off the stroboscope disc. (In the commercial type of stroboscope illustrated on page 174 the disc is replaced by a drum in which a number of holes are drilled. The neon lamp is seen through these holes as the drum rotates but again the "picture" afforded by the holes gives the speed at which the drum is rotating.)

PERCENTAGE INDICATOR

Another way of expressing the speed of an engine is in terms of a percentage of its maximum speed. The indicator used to show this value is similar in principle and operation to the normal a.c. indicator just described. It is driven by an a.c. generator to provide an indication of 100 per cent at maximum speed and will therefore show the appropriate percentage of maximum at lower engine speeds. The presentation on the dial is such as to shown one main pointer (Fig. 13.12) moving over a dial marked in units of 10 per cent, while a second pointer moves over a subsidiary dial marked in units of 1 per cent.

The percentage indicator is checked by means of a stroboscopic tester. The accuracy of the indication can be determined from the fact that it is arranged for the generator to run at 4,200 r.p.m. at maximum engine revolutions and, therefore, at 4,200 r.p.m. of the generator (as indicated by the stroboscope) the indicator should read 100 per cent.

FIG. 13.12. *Percentage-type engine-speed indicator.*

SYNCHROSCOPE

On a multi-engined aircraft it is necessary to run all engines as closely as possible to the same speed. This is sometimes difficult to achieve with the conventional indicators because of the separate tolerances on each indicator. In these circumstances a synchroscope is used. The synchroscope is an indicator which shows when two or more engines are running in synchronism. A two-engine synchroscope consists of a three-phase stator and a three-phase wound rotor. The power supplies to stator and rotor are derived from the two separate generators which are supplying the normal engine speed indicators. The speed indication and synchroscope for a twin-engined aircraft can be shown schematically as in Fig. 13.13.

Principle of Operation. If the stator of the synchroscope is receiving a three-phase supply say from the port generator, then a rotating field is created whose speed depends on the frequency of the generator supplying it. The effect of this rotating field is such as to provide a torque to move the rotor in the same direction as the field. However, the rotor is also receiving a three-phase supply from the starboard generator and the effect of this is to create a rotating magnetic field round the rotor (Fig. 13.14). The speed of this rotating field is proportional to the frequency of the starboard generator, while its effect is to produce a torque between rotor and stator such as will tend to rotate the rotor. If the field from the stator is in an anti-clockwise direction it will tend to move the rotor anti-clockwise. Should the field in the rotor be anti-clockwise the effect will be to drive the stator anti-clockwise. Since the stator is not free to move, then the torque will react to drive the rotor in the opposite direction.

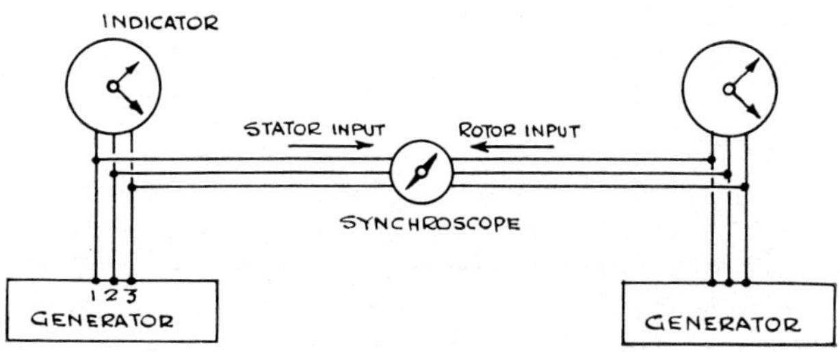

(*Above*) Fig. 13.13.

(*Below*) Fig. 13.14.

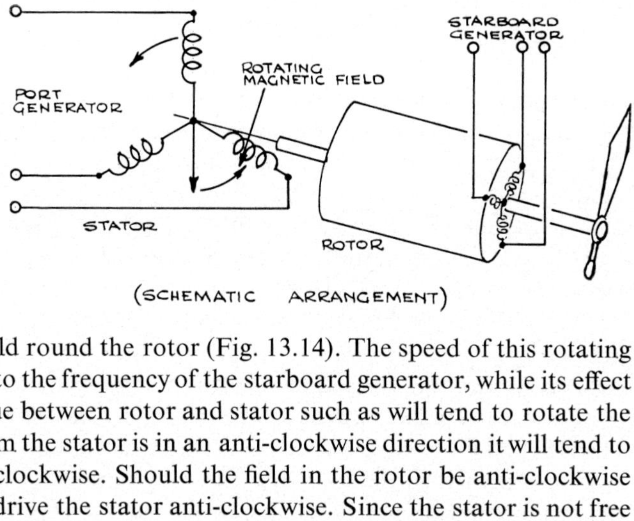

FIG. 13.15. *Kelvin four-engine synchroscope.*

Now, if the frequencies of the two generators are identical, then these opposing torques are identical and the rotor will be stationary. Should the speed of the starboard engine vary, the frequency of its generator will vary. Thus if the speed decreases, then the frequency decreases and the clockwise torque is reduced, so that the rotor will turn in the direction of the stronger torque; i.e. anti-clockwise. The reverse is true if the starboard engine speed should increase to a value greater than that of the port engine.

To summarize, the rotor is stationary when the starboard engine speed is the same as that of the port engine but will rotate clockwise or anti-clockwise if the starboard engine is faster or slower than the port engine. A pointer mounted on the rotor shaft will therefore rotate in one direction or the other to indicate the speed of the starboard engine relative to the port engine.

Four-engine Synchroscope. In a four-engine synchroscope (Fig. 13.15) there are three movements as just described. In each case the stators are connected to one engine generator and this is regarded as the master engine. The rotors are connected one to each of the generators on the other three engines. To achieve synchronization, the master engine is run at the required speed and the other engines are controlled so as to bring their respective pointers on the synchroscope to a standstill. The sensitivity of the device is such that the engines can be controlled to within 5 r.p.m. of the master.

Testing. In order to test a synchroscope it is necessary to have two variable-speed drives. One drive operates a generator, which is then connected to the stators of each of the synchroscopes. The other drive operates a second generator to which is connected each of the synchroscope rotors. In both cases the connections are parallel ones and normal engine-speed indicators are connected to the generators to form a load. The generator operating the stators is therefore to be regarded as the master, and the second generator speed is varied to bring the three pointers stationary. The speed difference between master and second generator should not be more than 5 r.p.m. The speeds for these checks are given in the manufacturer's specifications.

CHAPTER 14

Electrical Gyroscopes

THE principle and operation of gyroscopes for use as heading, turning and attitude indicators has been outlined in Chapter 9. This chapter also described the mechanical application of the principle in the directional gyroscope, artificial horizon and turn-and-slip indicator.

On aircraft flying at high altitudes it has become common practice to fit electrical gyroscopes. This is because the engine-driven vacuum pump, which provides the gyroscope drive for the mechanical instrument, becomes less effective as the altitude increases and the operation of the gyroscope suffers accordingly. Additionally, air-driven types suffer from the difficulty of ensuring clean air for driving purposes, whereas the electrical instrument can be housed in a hermetically sealed case.

Considering the directional gyroscope, it is usual to offer this as part of a complete compass system having magnetic detectors controlling the gyroscope through electronic amplifiers. Such compass systems demand more consideration than can be afforded them in this book and are therefore not included. The artificial horizon and turn-and-slip indicator are included, as both exist as separate electrical instruments.

ARTIFICIAL HORIZON

The principle and operation of the electrical artificial horizon is identical with that of its mechanical counterpart. Differences lie only in the fact that electrical equivalents of the mechanical actions are employed. The rotor is actually the rotor of a three-phase motor operated by a 400 c.p.s. supply at 115 volts. This produces a rotor speed of 12,000 r.p.m. to provide the degree of rigidity required in the gyroscopes.

Gyroscope Erection. As in the mechanical gyroscope, use is made of a gravity device to provide erection and to keep the spin axis vertical with respect to the earth. In this instance, the gravity device is usually a mercury switch (Fig. 14.1) which acts in the same way as a spirit level; i.e. if the gyroscope on which the mercury switch is mounted is not level, the mercury runs to one end of the tube containing it. As a result a circuit is completed through a small electric motor which is designed to create a torque between rotor and stator of the motor. The torque motor, on being energized, applies a torque to the gyroscope to provide erection until such time as the mercury switch reacts to the level condition and switches off the motor. There are two mercury switches each operating its own torque motor. One mercury switch is mounted on the gyroscope in the fore-and-aft sense to provide pitch erection. The other switch is mounted athwartships to provide roll erection. The circuit is shown in Fig. 14.1. It has an input of 115 volts at 400 c.p.s. to phases *A*, *B* and *C*. Colour indications are as follows: *BK*, black; *BU*, blue; *G*, green; *R*, red; *W*, white; *Y*, yellow.

Turn Errors. Because of the effect of centrifugal force on the athwartships mercury switch during a turn there will be a loss of datum of the indicator in roll. As the turn continues, the gyroscope converts the roll error into a pitch error. Thus, on coming out of a turn a slight roll and pitch error may be found on the indicator. In some indicators attempts are made to reduce this

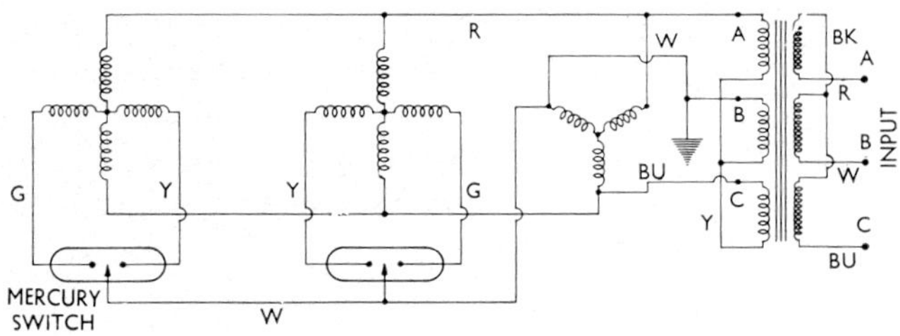

FIG. 14.1.

error by pivoting the gyroscope spin axis out of the vertical. In other cases it is possible, if the indicator is part of an integrated system, to provide monitoring by what is called pitch/bank erection.

Indication. The indication still makes use of a mechanical lever (Fig. 14.2) which carries the horizon bar and which is operated from the inner gimbal. Pitch and roll indications are the same as in the mechanical type, but it is usual to find that the electrical instrument has complete freedom in roll. In the mechanical instrument this is usually limited to about ± 115 deg.

Maintenance. Maintenance of the artificial horizon will, of course, depend on the type being considered. In general, however, the only test possible on the aircraft is to see that the horizon bar "settles" very closely to the datum position within $1\frac{1}{2}$ minutes of switching on. Other tests include erection (to determine the efficiency of the torque erection motors), roll pitch and yaw, coasting and power consumption tests, the details of these tests varying with the manufacture of the instrument.

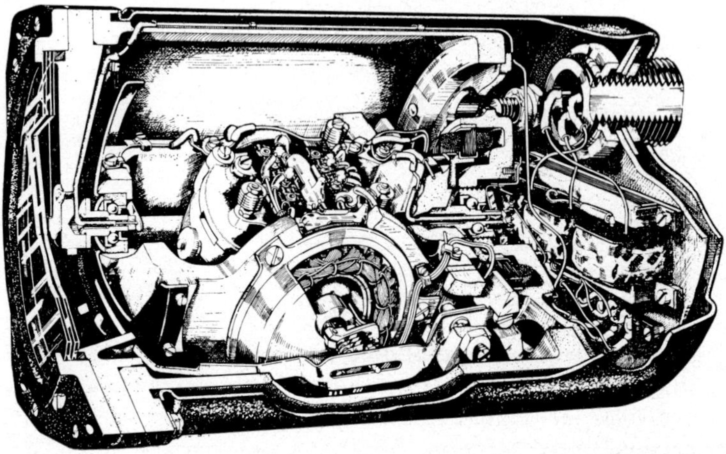

FIG. 14.2. *Gyroscopic indicator showing lever which carries horizon bar.*

TURN-AND-SLIP INDICATOR

The electrical turn-and-slip indicator is usually a d.c.-operated instrument where the gyroscope rotor is the rotor of a d.c. motor with a permanent magnetic field. This means that variations of d.c. voltage supply will produce changes of rotor speed. Compensation is achieved by a centrifugal governor (Fig. 14.3) which introduces resistance to reduce overspeed or removes resistance to counter reduced speeds. The rotor speed is thereby maintained reasonably constant at about 4,200 r.p.m.

The indications of rates of turn are the same in this instrument as in its

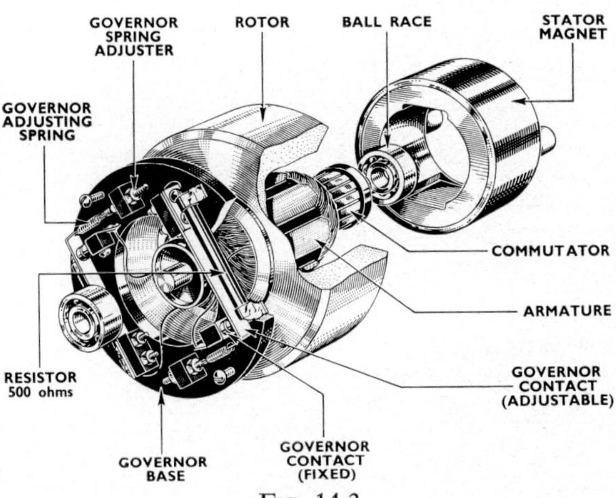

FIG. 14.3.

Table 14.1

CALIBRATION OF TURN INDICATOR

Dial marking	*Rate of Turn in degrees/ minute*	*Bank angle (degrees)*	*Tolerance*	
			Min:	*Max:*
1	180	35	174	186
2	360	55	342	378
3	540	69	472	608

mechanical counterpart and calibration is achieved in the same manner; i.e. by adjusting the strength of the control spring. To achieve dead-beat indication it is necessary to damp any oscillation of the gyroscope resulting from measurement of gyroscopic torque against a control spring. This is effected by an electro-magnetic device consisting of a copper drag-cup which is driven by the gyroscope gimbal as it moves to show the rate of turn. The drag-cup moves in the field of a permanent magnet and the resulting eddy-currents provide the required damping.

Included in this instrument is a failure indicator consisting of a flag marked OFF which appears in an aperture in the instrument dial. The flag is operated by a magnet which keeps it out of sight while the gyroscope wheel is rotating. When the wheel is stationary the flag magnet is attracted by the stator magnet to show the flag in the aperture.

The circuit of the turn indicator is shown in Fig. 14.4 where it will be noted that a suppressor unit is included. It is important to see that the polarity of the supply to the instrument is correct, otherwise the gyroscope wheel will run in the wrong direction, resulting in opposite turn indications. The slip indicator is of the ball type previously described; the complete instrument is shown in Fig. 14.5.

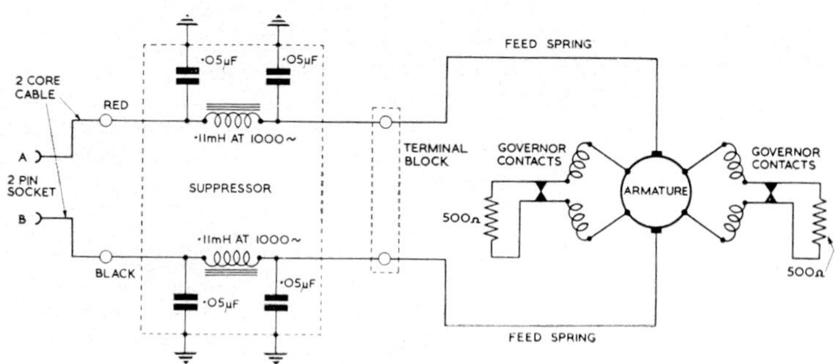

FIG. 14.4.

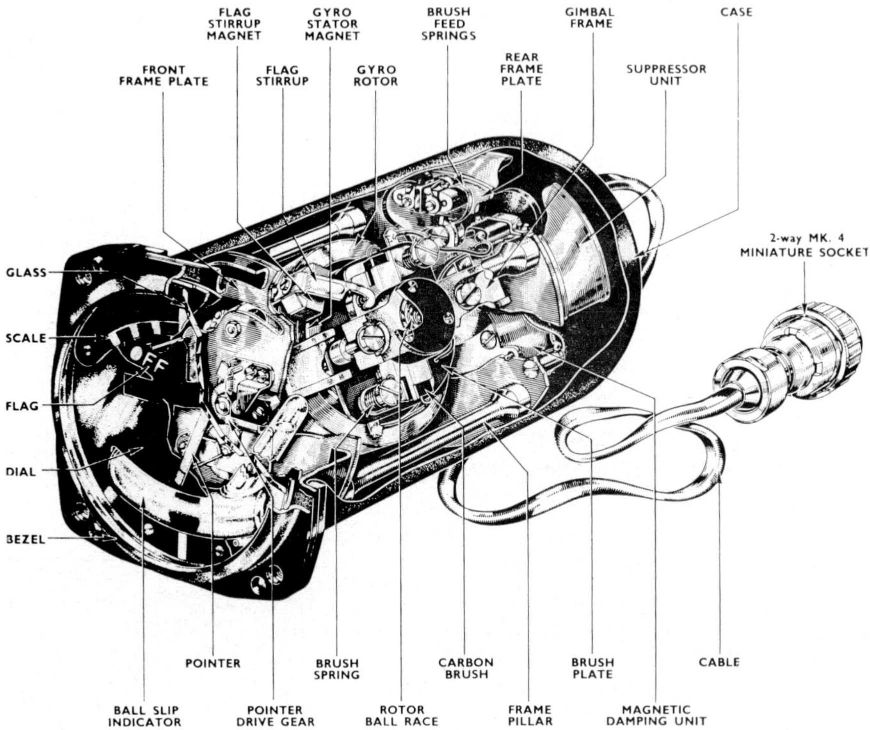

FIG. 14.5. *Turn-and-slip indicator.*

Calibration and Maintenance. The turn indicator shows scale marks for Rate 1, 2 and 3 turns. These can be checked by mounting the instrument on a turntable and checking according to the figures given in Table 14.1. The bank-angle figures refer to the angle to which the instrument should be banked on the turntable before starting the test. The instrument can be checked for signs of visual damage, and, on switching on the supply, the gyroscope rotor should be heard to gather speed and run smoothly. The OFF flag should move out of sight behind the dial when the power is switched on. Other tests will be carried out on a gyroscope test table, as already mentioned.

CHAPTER 15

Capacitance-type Fuel Gauges

In Chapter 7, mechanical fuel gauges were described and it was shown that these had disadvantages that could be summarized as follows:

(i) Changes of aircraft attitude could produce quite appreciable errors in the measured quantity.

(ii) The instruments are not of the remote-reading variety and are not suitable for the large modern airliner.

The latter disadvantage could be overcome with a remote indicating system such as the Desynn system described in Chapter 10. However, there still remains the disadvantage of attitude errors. Additionally, the modern airliner has such a number of fuel tanks that if the contents are read individually, some mental arithmetic on the part of the pilot or engineer is required to arrive at the total quantity of fuel at any one instant.

The capacitance type of fuel gauge is designed to minimize attitude errors and to provide a system capable of indicating the contents of a single tank or the sum total of fuel in a given number of tanks. The gauge is operated from the aircraft electrical supply and indications given by the gauge are independent of slight variations in the supply. A further advantage is that the capacitance type of fuel gauge can be calibrated in terms of "mass" of fuel rather than fuel quantity, if this should be desired. Although there are different makes of capacitance fuel gauges available, the principle employed is much the same in each case.

Principle. As the name suggests, the principle of the gauge lies in measuring the capacitance between two electrodes forming an integral part of the gauge, this capacitance depending on the amount of fuel contained in a given tank or tanks. *Capacitance* is an electrical property of a circuit, and a *capacitor* is a device which possesses this property in a measurable degree. A simple capacitor may be imagined as consisting of two metal plates (termed *electrodes*) separated by insulating material (termed the *dielectric*).

In such a capacitor, the capacitance is proportional to the area of the electrodes and inversely proportional to the distance separating them; it also depends on the nature of the dielectric. The electrical property of any dielectric is its *permittivity*, which is expressed by a *dielectric constant*. Thus, in two capacitors of identical dimensions but having different dielectrics, the value of capacitance will depend on the dielectric constant of the substance separating the electrodes.

Now, in the tank unit of a capacitance gauge both air and fuel are used as the dielectric. The dielectric constant of air is unity, and that of the various fuels is much higher, so that if a given capacitor has its dielectric changed from air to fuel a change in capacitance takes place. Consider two electrodes mounted in an empty fuel tank. Now if the tank is slowly filled with fuel, the air between the electrodes will be displaced by fuel and a change of capacitance occurs. Thus by measuring the change of capacitance, the change in fuel level will be known.

Attitude Errors. The attitude errors produced by the float-arm type of gauge can be minimized by placing two or more capacitors in the fuel tank

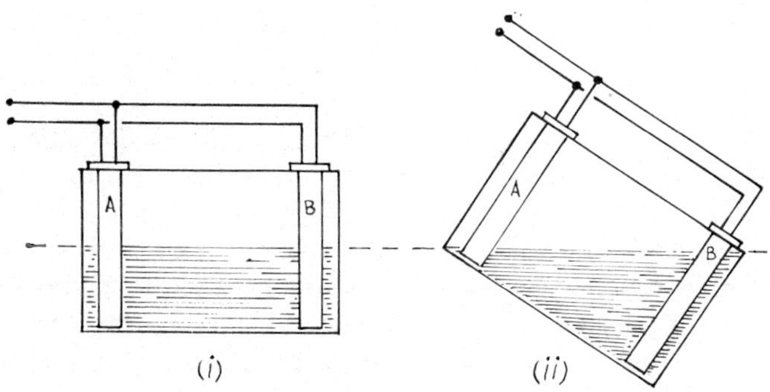

FIG. 15.1.

in appropriate positions (Fig. 15.1). In this instance two capacitors *A* and *B* are connected into a tank half filled with fuel. Thus each capacitor has an equal amount of fuel and air as a dielectric and the value of capacitance is equal to half the difference between the minimum and maximum of each. If the maximum value is regarded as unity and the minimum value zero then *A* and *B* each have a value of one half. By connecting the capacitors in parallel the output capacitance is the sum of the individual capacitors and the value of *A* plus the value of *B* is therefore one unit.

With the tank in position (ii) the fuel level at *A* has fallen to zero and therefore capacitor *A* is at minimum value but *B* will be seen to be at maximum value; i.e. unity. The capacitance of *A* plus *B* is therefore still equal to one unit and, ideally, no change of capacitance has occurred in spite of the large change in attitude of the tank. In practice, of course, it will not be

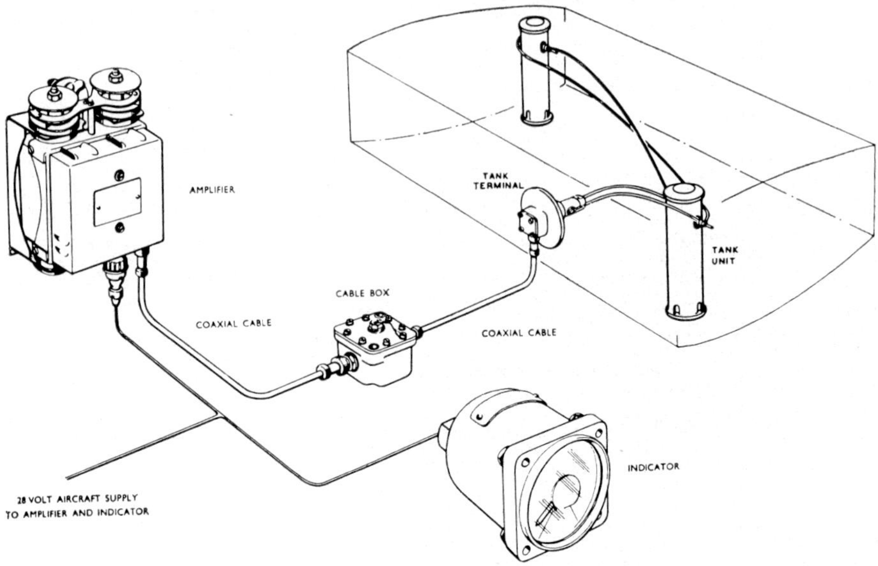

FIG. 15.2.

possible to effect such ideal conditions but, by carefully siting two or more capacitors according to tank shape, it is possible to minimize attitude errors.

Summation of Tanks. Since the measurement of fuel in the tank is really a measurement of capacitance, and capacitors connected in parallel provide a total capacitance equal to the sum of their individual values, then to measure the sum total of fuel in a number of tanks it is necessary only to switch the capacitors into a parallel network.

It would appear from the foregoing that the instrument which indicates the capacitance change could be calibrated in terms of fuel quantity. While this is so, it is also possible to measure the "mass" of fuel in the tank since the specific gravity of the fuel plays a part in determining the dielectric value of the fuel. In fact, the indicator calibrated in terms of units of mass provides a better reading than one calibrated in units of quantity. This can be explained in the following manner; consider a tank half full of fuel with two separate capacitance gauges connected to it. One such gauge is calibrated in units of quantity and one in units of mass. Let us assume that, at a given instant, both gauges are showing accurately the fuel state. Now consider an increase in ambient temperature which eventually will raise the temperature of the fuel in the tank causing it to increase in volume. Thus, although there is an increase in the "quantity" of fuel the mass of the fuel must remain the same.

What of the indicator readings? In fact, the increase in temperature which increased the volume of fuel also reduced the dielectric nature of the fuel. Thus, the capacitance value remains substantially the same and therefore the indicator readings do not change. Therefore, the indicator calibrated in mass units is still reading correctly, whereas the indicator calibrated in units of quantity *ought* to have increased its reading to be correct. Since the aircraft engine depends for its power on the calorific value of its fuel, which in turn depends on the mass of fuel, it is becoming common practice to calibrate capacitance-type fuel gauges in terms of mass rather than quantity.

SMITH-WAYMOUTH CAPACITANCE FUEL GAUGE

The principle of the Smith-Waymouth fuel gauge is as already described; i.e. the measurement of a change of capacitance of a capacitor with changing fuel in a tank or tanks. A basic system will comprise capacitance-type tank units, cable connection box, amplifier and indicator. The system can be shown as in Fig. 15.2. In this basic system there are two capacitance tank units connected in parallel by a rigid co-axial cable. Thus, at the tank terminals there will be capacitance proportional to the fuel in the tank.

A co-axial cable connects the tank terminals to a connecting box which houses the means of "trimming" the tank units to a standard capacitance figure. The cables from the connecting box now allow this capacitance to be fed to the input side of an amplifier which, in turn, supplies an output to the indicator containing a moving-coil milliammeter type of movement. The output (mA) from the amplifier is proportional to the fuel in the tank and therefore the indicator can be calibrated in terms of fuel mass or quantity as desired.

The circuit is shown in Fig. 15.3. The total capacitance of the tank units,

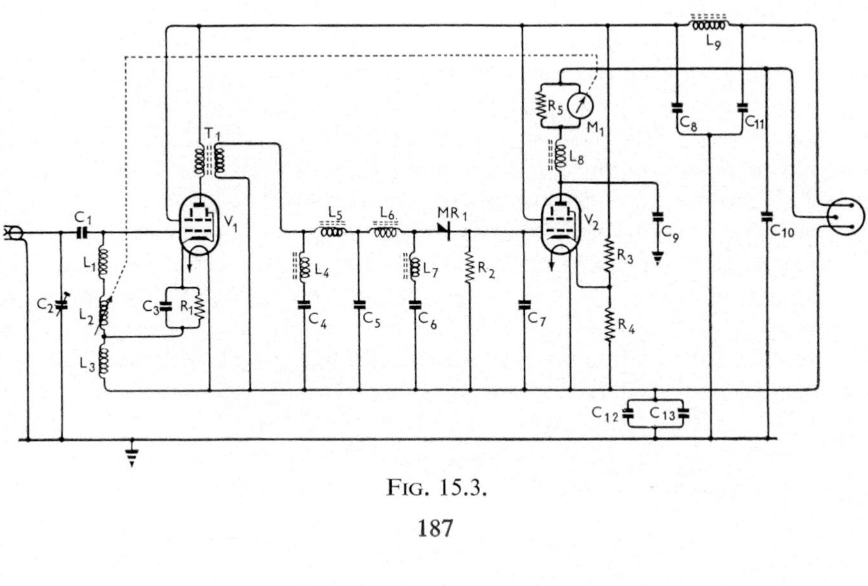

FIG. 15.3.

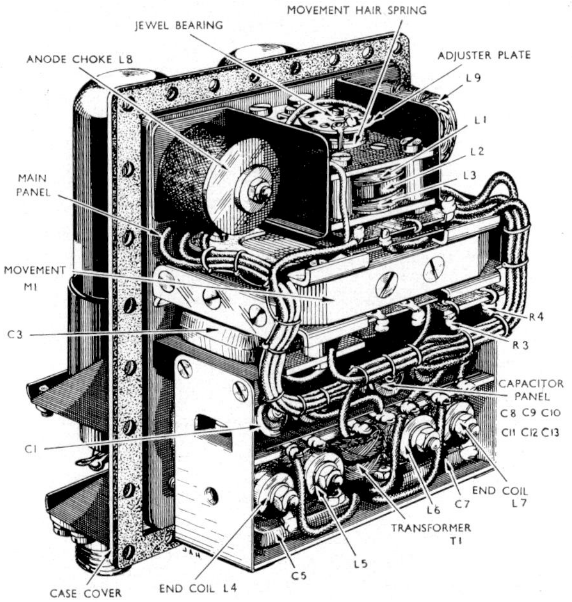

FIG. 15.4.

co-axial cable and connecting box is fed to the input side of the amplifier. Inductors L_1, L_2 and L_3 are connected across the capacitance input thus forming a circuit which oscillates at a frequency depending on the values of inductance and capacitance. If the inductance is regarded as being of a fixed value it will be appreciated that the oscillatory frequency is dependent on capacitance which, in turn, depends on the fuel contained in the tank. A valve, V_1, ensures that the circuit continues to oscillate at its natural frequency. The anode load of this valve (V_1) is the primary winding of the transformer T_1 the secondary winding of which feeds into a filter network consisting of inductors L_4, L_5, L_6 and L_7 and capacitors C_4, C_5 and C_6.

Now the output of this filter network depends on the frequency of the input from T_1 which, of course, is governed by the fuel in the tank. This output is fed via a rectifier MR_1 to a valve amplifier V_2 whose output is fed to the moving-coil milliammeter connected across the output terminals of the amplifier. The system is provided with a 28-volt d.c. supply from the aircraft and since this could vary to alter the output of the system, a feed-back circuit is incorporated in the amplifier to compensate for any voltage variation.

This feed-back system consists of a variometer movement M_1 forming part of the anode load of V_2. The purpose of the variometer is to vary the inductance of L_2 which forms part of the oscillatory circuit, and hence the true change of frequency of the oscillator is governed both by the change of capacitance input and by the change of inductance from the feed-back. The net result is an overall reduction in sensitivity but an independence of variations in the supply voltage. Two views of an amplifier unit are shown in Figs. 15.4 and 15.5.

Indicator. As mentioned previously, the indicator (Fig. 15.6) consists of a moving-coil milliammeter (Fig. 15.7). The pointer of this instrument covers

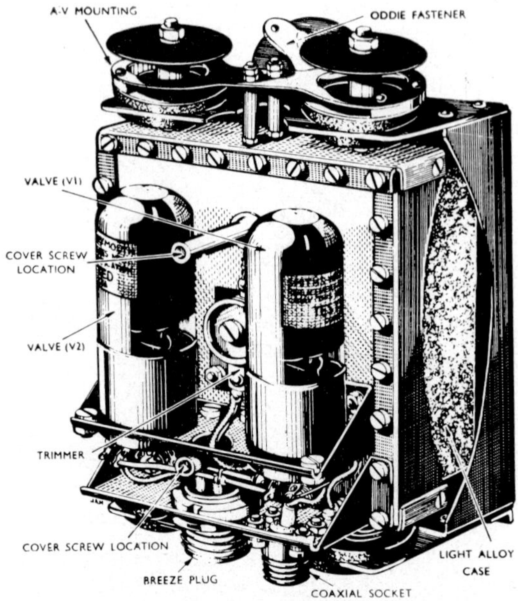

FIG. 15.5.

an arc of about 270 deg. and the maximum and minimum scale marks correspond to 2 mA and 7 mA respectively. The movement may be mounted in either a small or a large standard instrument case. A typical installation for a larger aircraft is shown in Fig. 15.8. This shows four tanks, with two tank units per tank, connected via cable boxes to four amplifiers feeding four separate indicators. Thus each indicator reads the contents of a single tank. Consider the right-hand half of Fig. 15.8, where it will be noted that in addition to the trimmer cable boxes there is also a cable box containing relays. Additionally, a selector switch is shown between the two indicators. The purpose of this arrangement is to overcome faults caused by an amplifier failure. This is done by the selector switch and relay, which can switch out of circuit the faulty amplifier and, at the same time, allow the remaining amplifier to indicate the contents of the tank in whose system the faulty amplifier lies. Cable boxes vary from installation to installation and two typical boxes are shown in Fig. 15.9. The tank units are simply capacitors which may be mounted in the tanks or which may even form part of the tank structure. The main types of tank unit are the "channel" type (Fig. 15.10), tubular type (Fig. 15.11) and flexible type (Fig. 15.12). This unit is mounted on a webbing strap.

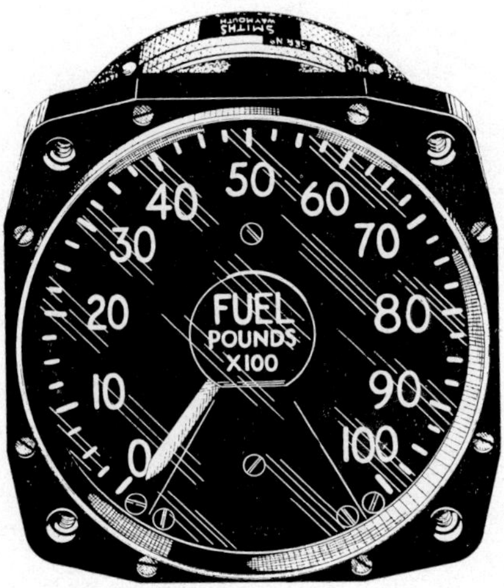

FIG. 15.6. *Indicator for Smith-Waymouth capacitance fuel gauge.*

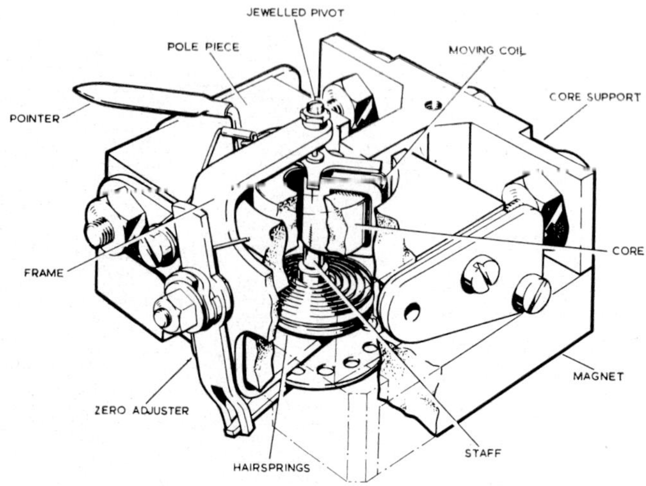

FIG. 15.7.

Maintenance. There is little direct maintenance to be carried out on this equipment other than a periodic functional check with a test set designed for the purpose. This set comprises a moving-coil instrument with suitable switching so that it can be used to indicate the supply voltage or the milliampere output of the amplifier, as desired. A variable capacitor is included in

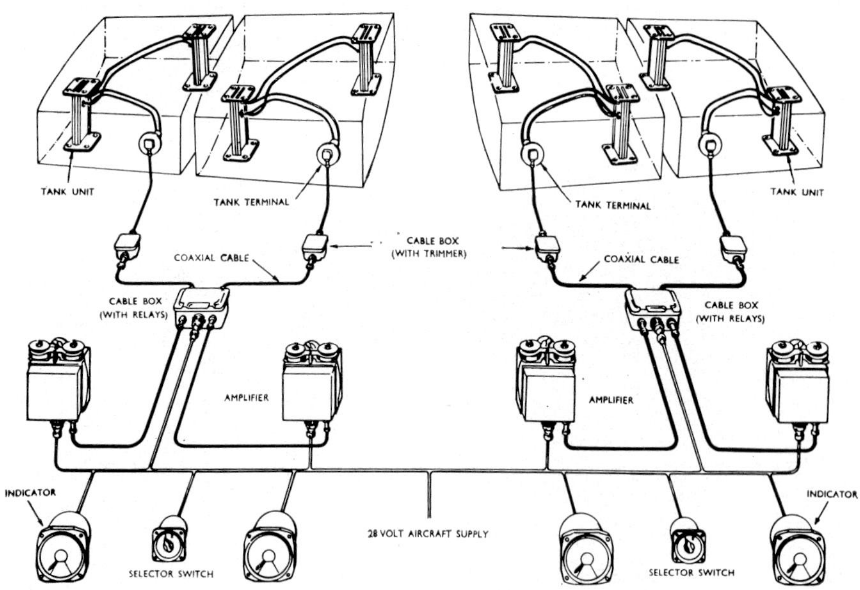

FIG. 15.8.

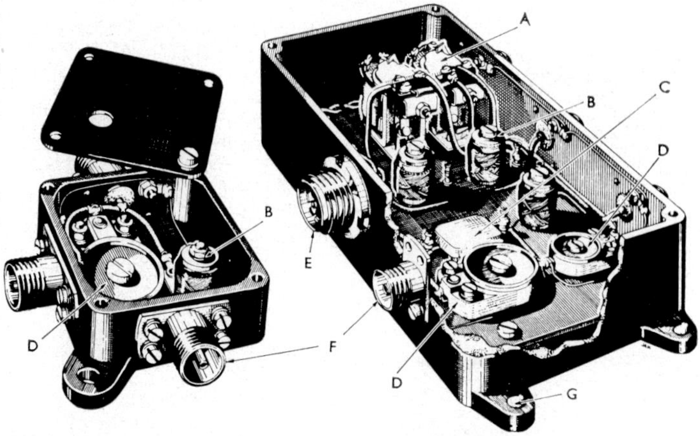

FIG. 15.9. *Unsealed cable boxes:* A, *relays;* B, *coil;* C, *fixed capacitor:* D, *trimmer;* E, *Breeze plug;* F. *co-axial cable;* G, *attachment bolt hole.*

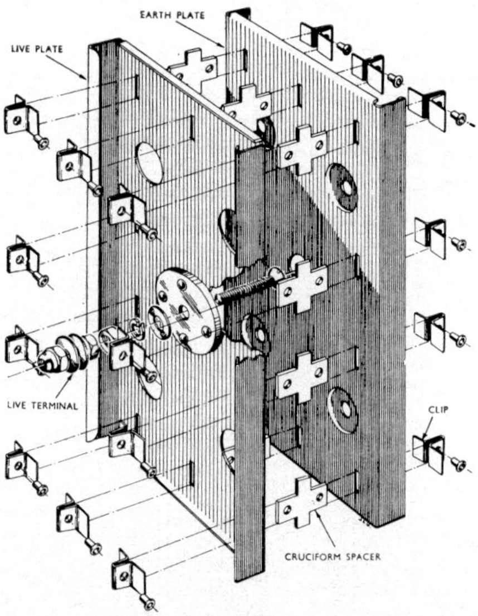

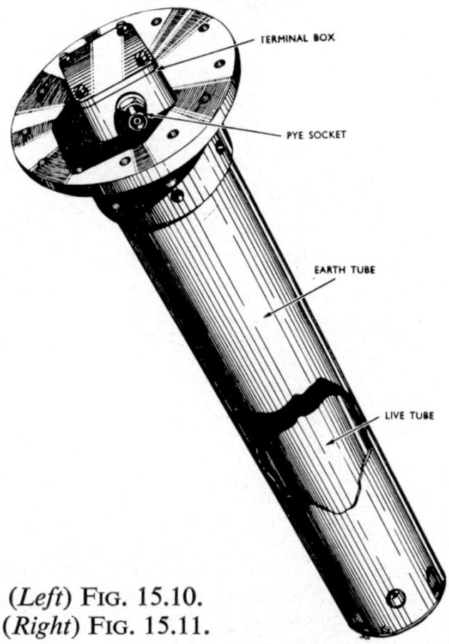

(*Left*) FIG. 15.10.
(*Right*) FIG. 15.11.

the test set and by means of special cables, also supplied, the set can be used to provide a changing capacitance input to the amplifier, thus simulating a fuel tank during refuelling. The test set is shown in Fig. 15.13, and full instructions for its use on any particular installation should be obtained from the manufacturers.

SIMMONDS PACITOR FUEL-CONTENTS GAUGE

The Simmonds Pacitor fuel-contents gauge also makes use of the principle of the variation of capacitance with changing fuel levels. In this

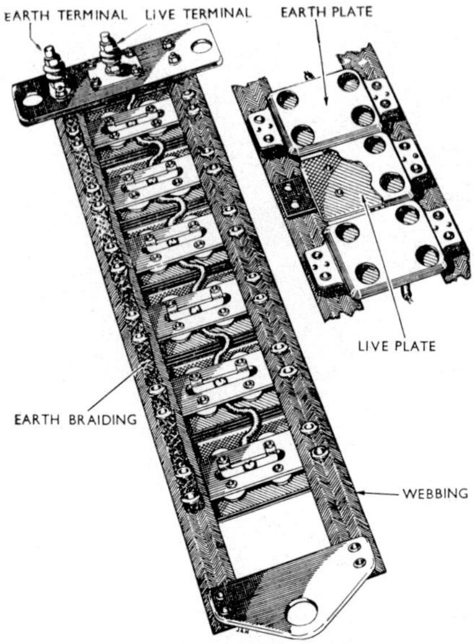

FIG. 15.12.

instance, however, the capacitance of the tank units is compared with a standard capacitor, the results being shown on a ratiometer-type movement. Thus the instrument is substantially independent of variations in the supply. The principle of the circuit is shown in Fig. 15.14 where it will be seen that the input is an alternating voltage. Two or more tank-unit capacitors are connected in parallel and govern the current through one ratiometer coil, while the standard capacitor governs the current through the second ratiometer coil; the resulting movement of the pointer is a measure of the ratio of currents in the two coils. In practice, the indicator houses transformers and rectifiers (to produce d.c. in the coils) and provision is made for adjustment of the indicator to the "empty" and "full" positions for calibration purposes. The basic system employs the following units: (*a*) tank units; (*b*) matching unit; (*c*) oscillator unit; (*d*) indicator.

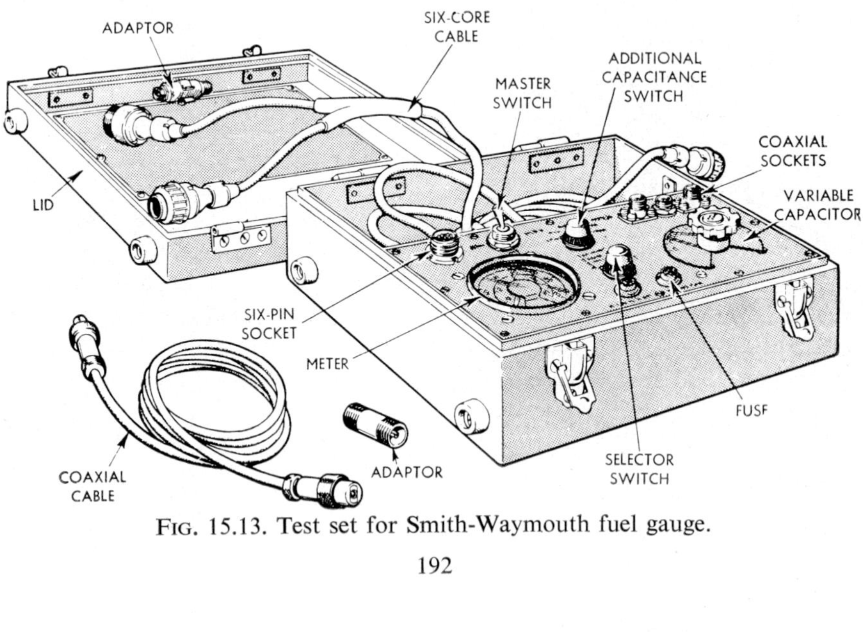

FIG. 15.13. Test set for Smith-Waymouth fuel gauge.

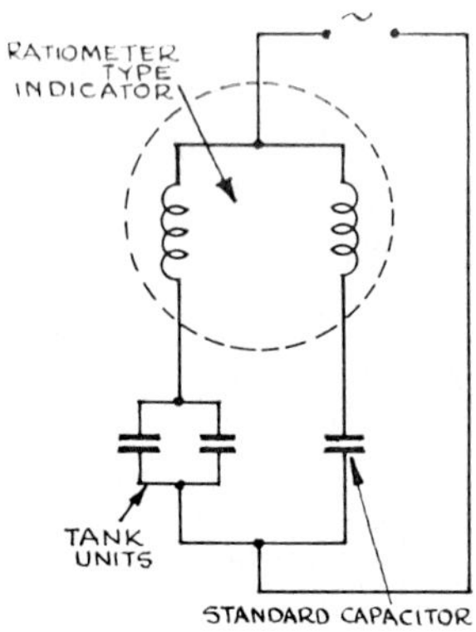

FIG. 15.14. *Circuit for the Simmonds Pacitor gauge.*

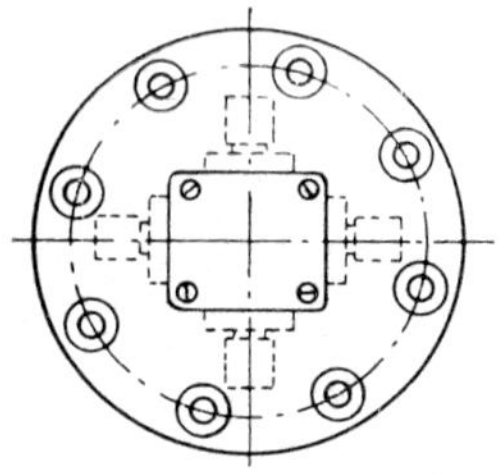

FIG. 15.15. *Drop-in tank unit (right). A cross-section of the unit is shown above.*

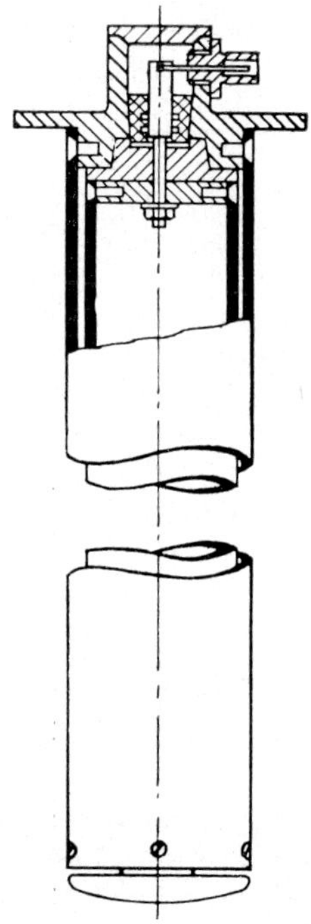

The tank units are, in fact, capacitors whose value depends on the fuel-tank contents. Tank units are of the "drop-in" type or the internally mounted type. The drop-in type (Fig. 15.15) consists of two concentric tubes arranged with an air gap between them, with one tube insulated from the other, thus forming a capacitor. Internally mounted tank units (Fig. 15.16) may also be of the tubular type but can be of a flexible nature to suit the tank construction. The internally mounted tank units are built into the fuel tank during manufacture.

The purpose of the matching unit is to receive the tank-unit output capacitance and compare this with a standard capacitance mounted in the unit. Its output is in the nature of two alternating voltages, proportional to the values of standard and tank capacitance, which are fed to the indicator. A schematic circuit of the matching unit is given in Fig. 15.17. The illustration shows that, by means of a co-axial cable connected to terminals *EF*, the tank-unit capacitance is added to the capacitance of a padding capacitor C_2; the standard capacitor C_1 has a value of 2400 pF. Terminals *C* and *D* carry a low-voltage input, at a frequency of $33\frac{1}{3}$ kc/s, to the two coupled circuits; the output terminals *A* and *B* supply the indicator.

The oscillator unit (Fig. 15.19) is the means whereby the standard aircraft supply of 115 volts, 400 c/s, is converted to a low voltage (approximately 9 volts) at a frequency of $33\frac{1}{3}$ kc/s. This oscillator output is supplied to the matching unit, as already described. The indicator (Fig. 15.20) is of the d.c. ratiometer type in which two coils, the deflection coil and the control coil, receive currents proportional to tank capacitance and standard capacitance respectively. A pointer traverses a dial to cover an arc of about 245 deg. to

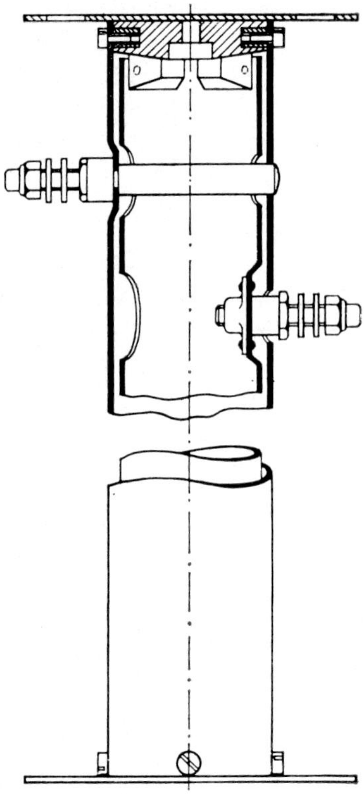

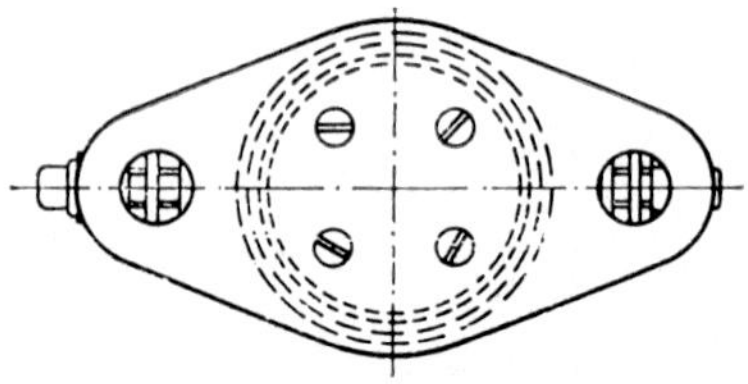

FIG. 15.16. *Internally mounted tubular tank unit. A cross-section of the unit is shown above.*

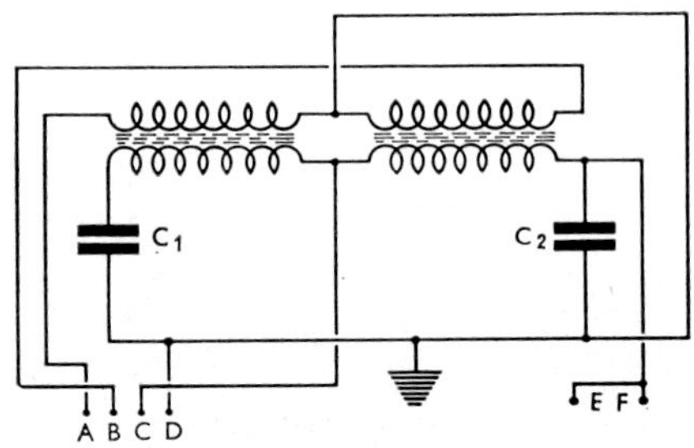

FIG. 15.17. *Schematic circuit of the matching unit.*

indicate the ratio of the currents and therefore the fuel-tank contents. Mounted in the instrument case are the necessary matching transformers and rectifier network to produce the direct currents from the alternating input of the matching unit. A resistance network included in the instrument provides a means of setting the "empty" and "full" positions of the pointer during the installation of the equipment in an aircraft. Calibration of the indicator may be in terms of quantity or mass according to requirements.

To provide a more comprehensive fuel-gauging system, additional components are available for use with the basic system. These components include a level-warning unit, cross-over switch unit, water-detector probe, and various switch units. The level-warning unit is a device connected in parallel with the matching unit and indicator. It is used to provide a visual or audible warning of a given level of fuel in the tank. The level-warning unit may be set for High Level (it will then be useful during refuelling), or Low Level, when it will operate at near tank-empty condition.

(*Opposite page.*) FIG. 15.18. *Typical tank-unit installation.*

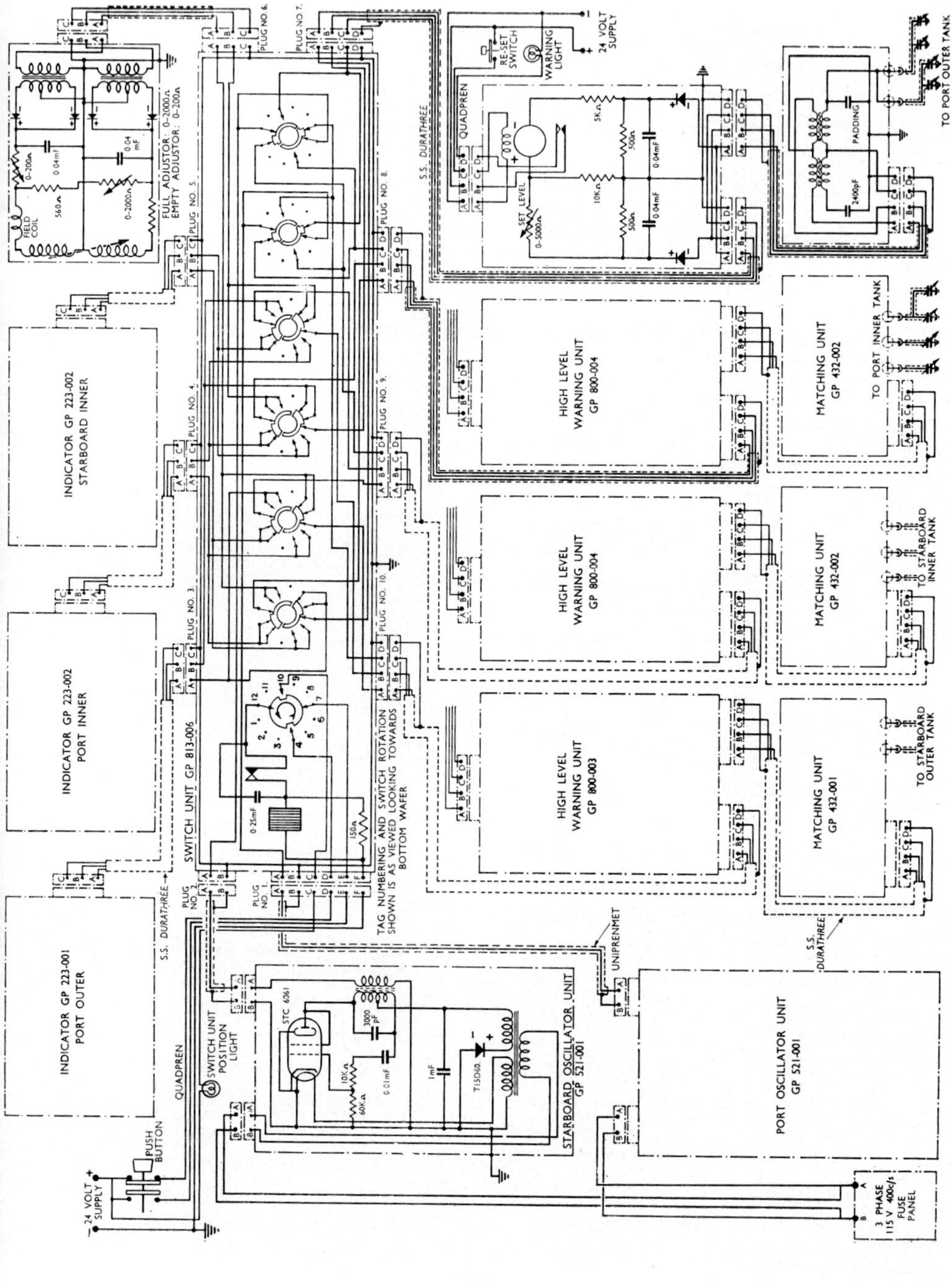
INDICATOR GP 223-001 PORT OUTER
INDICATOR GP 223-002 PORT INNER
INDICATOR GP 223-002 STARBOARD INNER
FULL ADJUSTOR: 0-2000Ω
EMPTY ADJUSTOR: 0-200Ω
FIELD COIL
SWITCH UNIT GP 813-006
TAG NUMBERING AND SWITCH ROTATION SHOWN IS AS VIEWED LOOKING TOWARDS BOTTOM WAFER
HIGH LEVEL WARNING UNIT GP 800-004
HIGH LEVEL WARNING UNIT GP 800-004
HIGH LEVEL WARNING UNIT GP 800-003
MATCHING UNIT GP 432-002
MATCHING UNIT GP 432-002
MATCHING UNIT GP 432-001
TO PORT OUTER TANK
TO PORT INNER TANK
TO STARBOARD INNER TANK
TO STARBOARD OUTER TANK
PADDING
QUADPREN
RE-SET SWITCH
WARNING LIGHT
24 VOLT SUPPLY
SET LEVEL
S.S. DURATHREE
UNIPRENMET
STARBOARD OSCILLATOR UNIT GP 521-001
PORT OSCILLATOR UNIT GP 521-001
SWITCH UNIT POSITION LIGHT
PUSH BUTTON
24 VOLT SUPPLY
3 PHASE 115 V 400c/s FUSE PANEL

FIG. 15.19. *Simmonds oscillator for use with the Pacitor system.*

(*Below*). FIG. 15.20. *Ratiometer-type indicator for use with the tank unit.*

The cross-over switch unit is used in larger installations where there are several Pacitor systems. It is used during an emergency in the event of a failure of any one part of a system, and enables the contents of the tanks in the faulty system to be read on the indicator of another system. This cross-over switching operation is performed by push-button control. The water-detector probe (if fitted to a Pacitor system) indicates the presence of water which has accumulated in the fuel tank. The probe is push-button operated and indicates this presence by suppressing the appropriate indicator pointer to a below-zero reading. In the absence of water in the tank, an approximate mid-scale reading is obtained. The switch units are distinct from the cross-over switch units and can be provided where it is required to indicate the contents of separate tanks in turn, and the sum total of all fuel if necessary. A typical installation is that of the Britannia aircraft, which comprises tank units, matching units, high-level warning units, cross-over switch unit, oscillators and indicators. The schematic circuit of this installation is shown in Fig. 15.18.

Maintenance and Testing. Little routine maintenance is required on the Pacitor system apart from a periodic check of the output and frequency of the oscillator unit. Testing is carried out, using the appropriate Pacitor Test Set, and details of maintenance and testing should be obtained from the manufacturer, who can also supply useful "trouble-shooting" charts.

Part III

TEST PROCEDURE

CHAPTER 16

Testing and Test Equipment

AIRCRAFT instruments are subjected to periodic calibration checks as recommended by the A.R.B. in conjunction with the particular instrument manufacturer. These tests are such as to satisfy the requirements that the instrument will "adequately and reliably perform the functions prescribed". (See Chapter 1: *Mandatory Equipment*). To help facilitate these tests numerous standard items of equipment are available, and some of the more common ones in use to-day are described in the following pages.

Errors in Testing. In spite of the high quality of the various items of test gear available, many instruments can be rejected as being outside their particular tolerance, not because they are at fault, but owing to faulty application of the test equipment by the operator. It is difficult to be specific about the whole range of testing but the operator may care to bear in mind the following points before commencing a test: (1) When reading an instrument it is essential that the eye of the operator be in line with the pivot of the pointer on the instrument under test; this ensures a minimum parallax error. (2) In connecting an electrical instrument to a test set see that all contacts are clean, thus ensuring mimimum contact resistance. (3) Remember that the resistance of the connecting leads may materially affect the results obtained. (4) Check the voltage and frequency of any supply to the test set before commencing the test, and monitor this information if the test takes more than a few minutes. (5) With thermo-couple instruments remember that small thermal e.m.f.s may be generated at contact points owing to temperature variations. It should also be noted that the initial reading of the instrument, before the test begins, is the reading of the test-room temperature. Thus, the instrument should have been in the test room some minutes before the test begins.

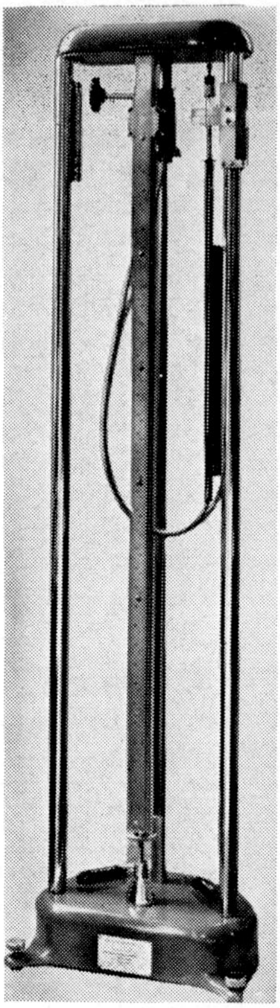

FIG. 16.1. *Bryans Master Manometer designed for the testing of low-pressure gauges.*

Mechanical Instruments. Many mechanical instruments are of the pressure-operated type and, in this case, it is essential to see that the connections between the indicator and the test set are pressure-tight; i.e. have no leaks. Additionally, care should be taken to see that the instrument cannot be overloaded by inadvertent operation of the pressure tester, otherwise the pressure-sensitive element may be strained beyond its elastic limit. When testing altimeters remember to make the appropriate allowance for altitude, temperature and latitude (see page 83). In the case of rate-of-climb indicators, the instrument should not be handled too much during testing unless it has a thermal shield around it, otherwise small temperature changes may materially affect the reading. Gyroscopic instruments of the air-driven type should have a supply of clean air, free from moisture, to operate them.

PRESSURE-GAUGE TESTING

In Chapter 3, the principle of checking pressure gauges was discussed and it was seen that low-pressure gauges were tested with a U tube or manometer, while high-pressure gauges were tested with a dead-weight tester. The practical forms of these pressure-gauge testers are more refined than the simple illustrations previously given and are described in this chapter.

Low-pressure Manometer. A typical manometer for low-pressure gauge testing is the Bryans Master Manometer (Fig. 16.1). This is designed for checking low-pressure instruments to a maximum of 35 in. Hg, and for suction instruments the scale is calibrated from 32 to 0 in. Hg. It consists of a mercury cistern (see Fig. 16.2) which is connected by a flexible tube to a short length of glass tubing on which is marked a datum line. The cistern can be moved up and down a central scale, final adjustment being made by means of a vernier. With the cistern set at zero on the scale, a float in the glass tube will rest against the datum mark. As the cistern is lowered down the scale, the float will drop in the glass tube. If pressure is now applied to the cistern, the mercury contained therein will be pushed up the flexible tube and so raise the float. By adjusting the cistern with the vernier until the float is again showing against the datum mark on the glass tube, the pressure can be read off

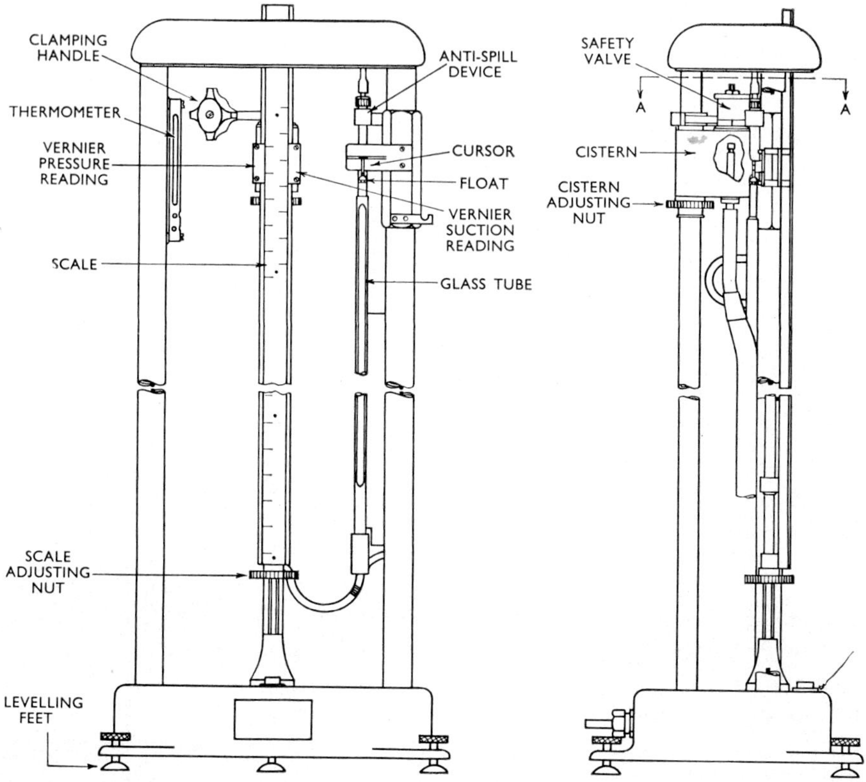

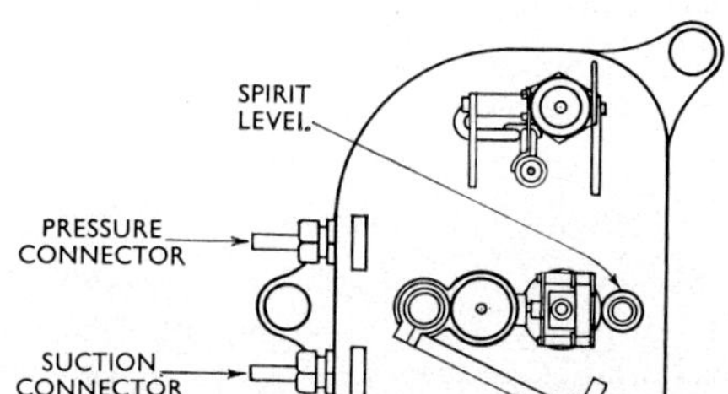

FIG. 16.2. *The main features of the Bryans Master Manometer shown in* FIG. 16.1.

from the position of the cistern on the scale.

The principle is shown in Fig. 16.3. In (*a*) it is seen that the float is against the datum mark and the cistern is at zero on the scale. (It is necessary to check the zero position from time to time by setting the cistern until the float shows against the datum mark and adjusting the zero device on the scale.) In (*b*) the cistern has been lowered and pressure has been applied to the system to raise the level of the float. When the float is resting against the

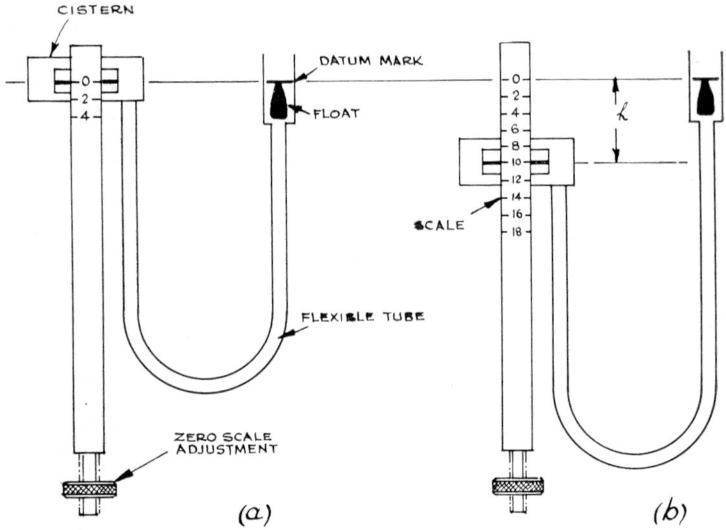

FIG. 16.3.

datum mark the pressure is equal to the difference, *h*, in levels between float and cistern. This difference in levels can be read directly off the scale, which is calibrated in inches Hg. Two connections, one pressure and one suction, at the back of the manometer base enable the instrument under test to be connected to the manometer. Thus direct comparisons can be made between the indicator under test and the manometer scale.

Variants of the Master Manometer. The manometer described has special variants. One of these is the Micromanometer which is similar in construction to the Master Manometer but uses oil as the liquid medium although the scale is calibrated in "inches of water". The range of the Micromanometer is 0–28 inches of water and some idea of its sensitivity can be gained from the fact that readings can be taken to within 0·005 inch of water. Another example is the manometer for the calibration of airspeed indicators. This has two parallel glass tubes (Fig. 16.4) one containing water and the other mercury. A switch enables one or the other to be selected and when pressure is applied the liquid level increases in the side chosen. A cursor is adjusted until it is level with the float in the glass tube and the pressure is read off the scale against the cursor. The range of this manometer is 0–900 m.p.h. (0–800 knots) although the scale can be varied to suit particular requirements; for example millimetres; millimetres and knots; millimetres and kilometre/hr, and so on.

High-pressure Gauge Tester. The dead-weight tester was mentioned as being used for checking higher pressures than can be measured with a manometer. A typical oil-filled dead-weight tester is shown in Fig. 16.5. This

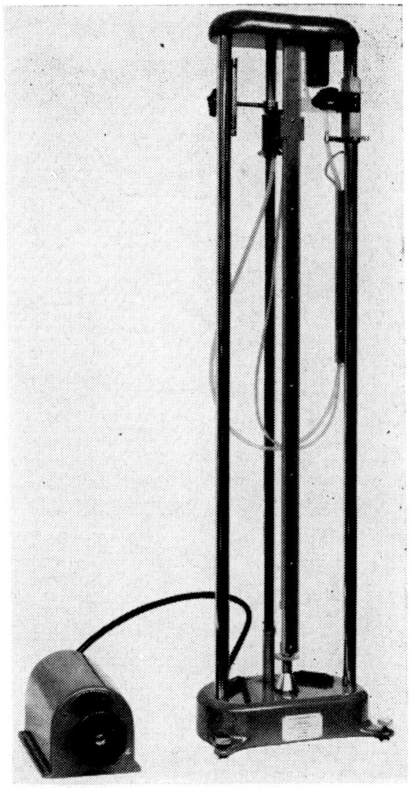

FIG. 16.4. *Bryans Micromanometer having a range of 0–900 m.p.h.*

has a dual range: 10–500 p.s.i.; 100–5,000 p.s.i. The same weights are used for either range but the pistons, on which the weights rest, are different in area. A weight carrier mounted on the piston enables the weights to be rotated to reduce "stiction" as pressure is applied. The weight carrier forms part of the total load; i.e. on the low-pressure side the weight carrier alone is equivalent to 10 p.s.i. while on the high-pressure side it is equivalent to 100 p.s.i.

A useful accessory to this tester is a pressure transmitter for oxygen-pressure gauges. Oil should not be permitted near an oxygen-pressure gauge, so it is not possible to connect the gauge directly to the dead-weight tester. The transmitter is therefore connected between the gauge and the dead-weight tester (Fig. 16.6). The transmitter consists of two chambers separated by a bellows unit. The lower chamber is connected directly to the tester, while the upper chamber is filled with carbon tetrachloride. When the dead-weight tester is operated, the oil pressure acts on the bellows and is transmitted to the carbon tetrachloride and thence to the oxygen-pressure gauge. Thus there is no contact between the gauge and the oil from the tester.

Leak Tester. On page 58 a brief description was given of the method of testing the pitot/static system of an aircraft. A simple tester, complete with various adaptors, is made by Bryans Aeroquipment Ltd. The leak tester (Fig. 16.7) consists of a metal case which houses items (i) to (iv) now to be described. Item (i) is a three-way selector valve operated by a control knob on the front of the case; this knob can be set to select: (*a*) Pressure to Pitot; (*b*) Suction to Static; (*c*) Release to Atmosphere. A choke comes into operation during suction release to protect the rate-of-climb indicator from an excessive change of pressure. (It is assumed that the instrument panel is still connected to the pitot/static lines when using this tester.)

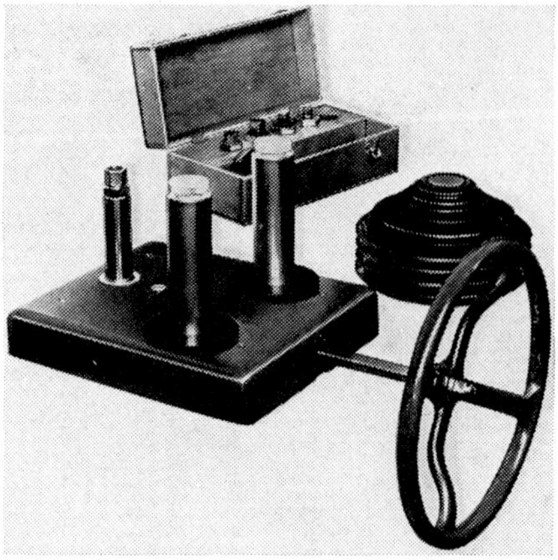

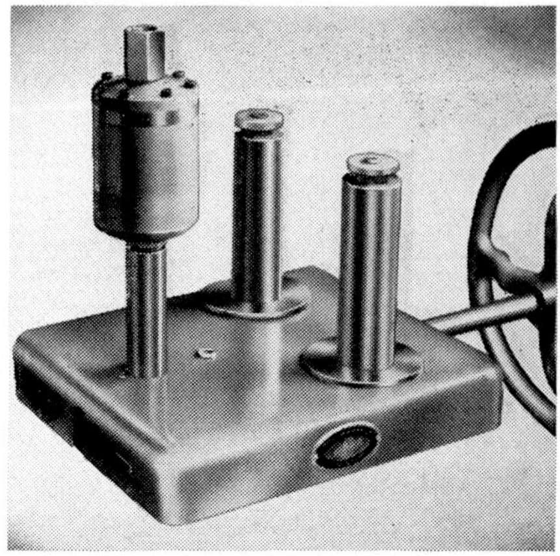

(*Left*) FIG. 16.5. *Oil-filled dead-weight tester.* (*Right*) FIG. 16.6. *Dead-weight tester and bellows unit for use with oxygen-pressure gauges.*

Item (ii) is a double-action non-return pump to enable negative or positive pressures to be fed to the selector valve. Item (iii) is an indicator having a segment covering a range of 115 to 130 knots. Any leakage in the system results in the pointer moving backwards over this range, and the time taken to measure this range is a measure of the leakage factor. Item (iv) comprises static and pitot connectors, fitted with replacement filters, to enable the tester to be connected by rubber hose to pressure-head or static-vent adaptors. Captive blanking caps on the tester are used to keep the system free from dirt when not in use, and enable a leak test to be made on the tester prior to use. Before operation, the leak tester itself should be checked as mentioned previously. It is then ready to test the aircraft pitot/static system by means of the appropriate adaptor. Various types of adaptor are shown in Fig. 16.8. The adaptors are used with pressure heads and static vents as indicated in Table 16.1.

FIG. 16.7. *Leak tester for pitot/static system.*

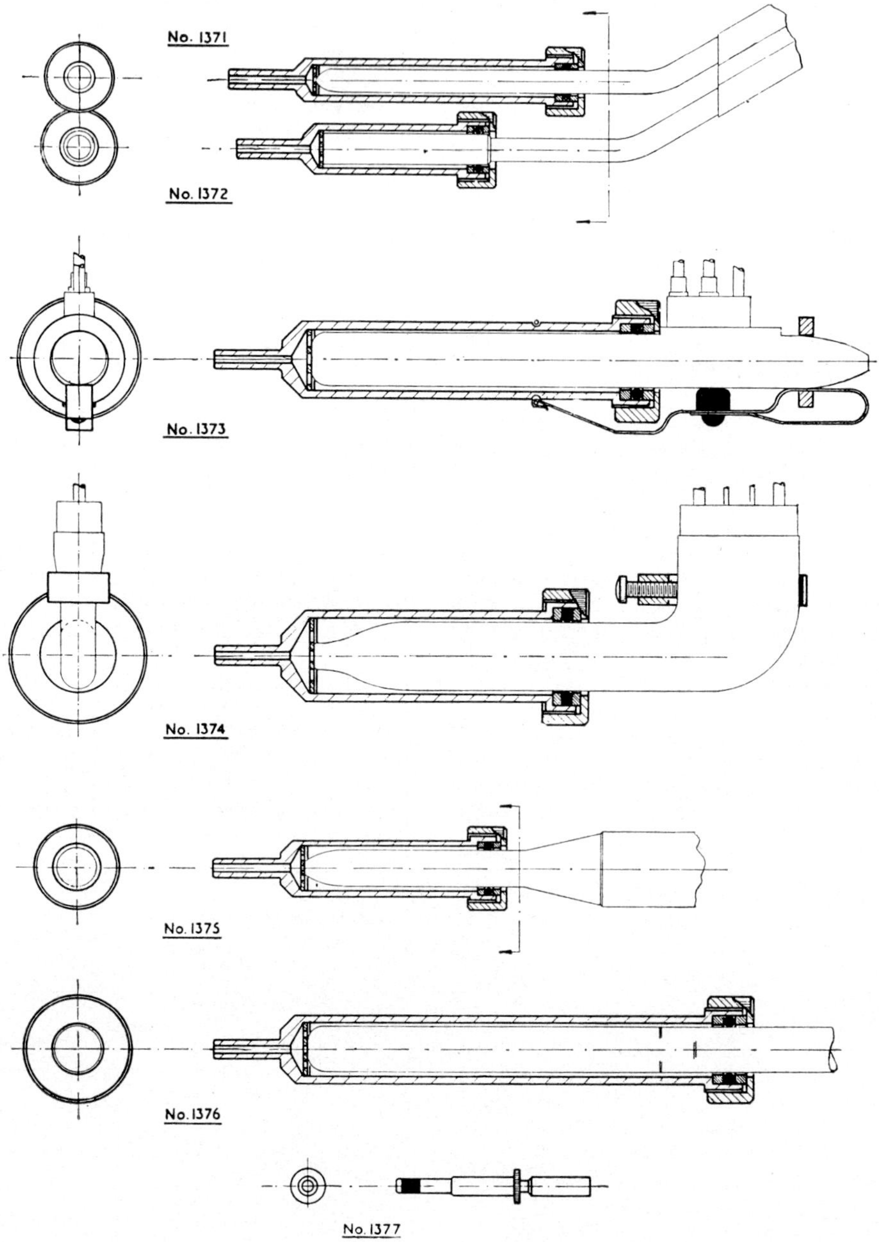

FIG. 16.8. *Various types of adaptors used with pressure heads and static vents.*

Table 16.1

ADAPTORS FOR PRESSURE HEADS AND STATIC VENTS

Adaptor No.	*Pressure Heads*			
	A.M. Ref. No.	*Type No.*	*American Ref. No.*	*American Spec.*
1371	6A/432	Mk. VIIA (Static Tube)	—	—
1372	106A/1438	Kollsman	AN.5815–1	AN–GG–T–791
		Kollsman	Army Type B–6	94—27716
		Mk. VIIA (Pitot Tube)	—	—
1373	6A/728	Mk. VIIIA	—	—
	6A/729	Mk. VIIIB	—	—
	6A/730	Mk. VIIIC	—	—
	6A/731	Mk. VIIID	—	—
	6A/874	Mk. VIIIE	—	—
	6A/875	Mk. VIIIF	—	—
	6A/1681	Mk. VIIIG	—	—
	6A/1682	Mk. VIIIH	—	—
	6A/1683	Mk. VIIIJ	—	—
	6A/1684	Mk. VIIIK	—	—
	6A/1861	Mk. VIIIM	—	—
	6A/2176	Mk. VIIIM	—	—
		Pioneer	AN.5814–1 and 2 (round section)	AN–T–1
	106A/636 & 106A/334	Pioneer	AN.5816–1 and 2 (round section)	AN–T–1
		Pioneer	AN.5810	AN–GG–T–791
		Pioneer	1079–S940J	—
1374		Kollsman	AN.5814–1 and 2 (rectangular sec.)	—
				AN–T–1
	6A/865	Mk. VIIIC	—	—
	6A/863	Mk. VIIIA	—	—
	106A/636 & 106A/334	Kollsman	AN.5861–1 and 2 (rectangular sec.)	AN–T–1
		372E–02	—	—
	106A/916	—	D–2	94–27874–B
1375	106A/1329	G–2	AN.5812–1	AN–T–44
	—	G–1	AN.5813–1	AN–T–44
1376	—	Mk. IX	—	—
	Static Vents			
1377	6A/1421	C	—	—
	6A/1508	D	—	—
	6A/1511	E	—	—
	6A/1686	F	—	—

EXTRA ADAPTORS AVAILABLE

1468	Static vent plug with warning flag. For use on aircraft with two static vents of the type listed above.
1469	Pipe-line adaptor for connecting the leak tester to the pitot/static pipe-line when the pressure leads on static vents are removed.

RATE-OF-CLIMB INDICATOR TESTS

The rate-of-climb indicator (page 85) was shown to consist of two main parts: (i) pressure gauge; (ii) calibrated leak. These separate assemblies are tested as such before final assembly and testing. Thus, there are three stages of testing, and a separate tester is available for each stage of the test; they will be discussed under three headings: (i) movement calibrator; (ii) orifice comparator; (iii) rate-of-climb tester (*a*) and (*b*).

Movement Calibrator. The movement calibrator consists of an inclined manometer for measuring the low pressures applied to the capsule of the rate-of-climb indicator. The inclined manometer (Fig. 16.9) is mounted on a metal base-plate and has three adjustable levelling feet which are used in conjunction with two spirit levels carried on top of the base-plate. A panel assembly carries a cistern, overflow container and associated piped connections. On the front of the panel are the instrument connections, the cistern adjustment knob and the cistern filling screw. A further connection allows a bellows control unit to be connected to the inclined manometer to provide the small positive and negative pressures required in this form of testing. A scale on the inclined manometer allows pressures of ± 115 mm. (water gauge) to be measured.

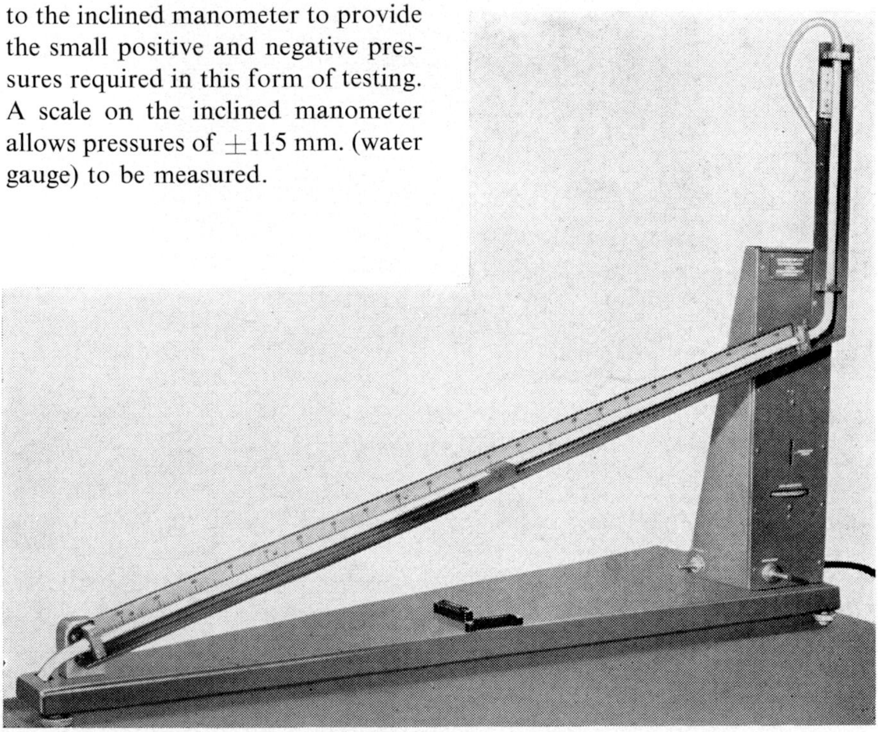

FIG. 16.9. *Rate-of-climb indicator-movement calibrator with inclined manometer. Spirit levels are shown at the centre of the baseplate.*

Although the scale is calibrated in water-gauge units the actual filling liquid is Duckham's Fluid Q.5228. The specific gravity of this fluid is 0·823 at 60 deg. F. and this is considered when making the scale. Should the temperature in the test room be different from this, corrections to the readings are made by a factor of 0·01225 per Fahrenheit degree; e.g. for temperatures above 60 deg. F., the reading is "contracted" by a factor of 0·01225 per degree, and for temperatures below 60 deg. F. the reading is "expanded" by the same factor. In operation, the following procedure should be observed. (i) Use the levelling screws and spirit levels to set up the tester. (ii) Operate the cistern adjustment knob to set the meniscus of the fluid at zero. (iii) Connect the

FIG. 16.10. *The orifice comparator, or calibrated-leak comparator, with flow-control valve and holder for standard leak.*

instrument under test and apply the pressure appropriate to the particular make of instrument under test.

Orifice Comparator. The orifice comparator or, more properly, the calibrated-leak comparator, is used in conjunction with a standard leak assembly provided by the manufacturer of the instrument under test. It consists of a multiple manometer (Fig. 16.10) mounted on a panel the base-plate of which has the usual levelling screws and spirit levels. A flow-control valve and a holder for the "standard leak" are carried on the front of the panel, and provision is made for connecting the leak assembly under test.

FIG. 16.11. *Bryans rate-of-climb tester, earlier model, fitted with synchronous clock.*

An overflow cistern is carried on the back of the panel and a viewing hole allows the amount of overflow liquid to be observed.

The multiple manometer comprises two vertical tubes and one inclined manometer tube which are supplied from a cistern common to all three tubes. The whole of this system contains distilled water. One vertical tube is open to the atmosphere, while the other vertical tube is connected to the input side of the standard leak. The inclined manometer tube is connected between the standard-leak output and the input side of the leak under test. The arrangement is such that when the leak under test is identical with the standard leak, the centre manometer shows a zero reading. In operation, the following procedure should be observed. (i) Level the comparator using the levelling screws and spirit levels. (ii) Connect the standard leak, appropriate to the leak under test, into the leak-holder on the front of the panel with the cap well sealed down. (iii) Connect the leak under test. (iv) Operate the control valve on the front panel until a reading on the inclined manometer, appropriate to the leak under test, is obtained. (v) Check the centre manometer, which will be at zero if the leak under test is identical with the standard leak. Any discrepancy between the leaks is shown as a percentage error, the range of error measured being ± 10 per cent.

Rate-of-climb Tester (*a*). Having used the testers—as described in the preceding paragraphs—to verify the accuracy of the component parts of the rate-of-climb indicator, another method of test is required after final assembly of these components. This is the rate-of-climb tester (Fig. 16.11) which verifies the accuracy of the completed instrument. On page 88 it was mentioned that the rate-of-climb indicator could be tested, using a barometer and a stop-watch. In effect, the rate-of-climb tester provides both these facilities in one unit. The tester consists of a mercury manometer, fitted with electrical contacts, and a synchronous clock with a dial marked in seconds.

The manometer comprises a metal cistern with a precision-bore glass tube. An insulated cap at the top of the tube carries three contact rods of different lengths which are suspended in the tube. The difference in length between rods 1 and 2 is such that when mercury in the tube is rising at a rate proportional to 1,000 ft./min. climb it will take exactly one minute to rise from the end of rod 1 to the end of rod 2. Similarly, the difference between rod 1 and rod 3 is such that a rate of climb of 3,000 ft./min. will cause the mercury to rise from the end of rod 1 to the end of rod 3 in exactly one minute. The synchronous clock provides a means of measuring the time interval between rods when the mercury is rising, or falling, at a rate proportional to 1,000 or

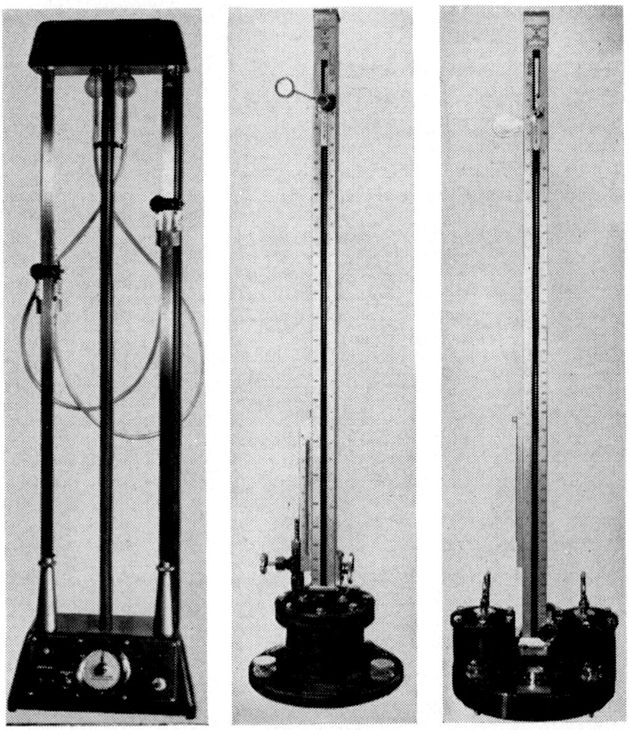

(*Left to right*). FIG. 16.12. *Bryans rate-of-climb tester*. FIG. 16.13. *Negretti & Zambra standard test barometer*. FIG. 16.14. *Negretti & Zambra twin-cistern barometer*.

3,000 ft./min. A selector switch on the front panel allows the tester to be set for 1,000 to 2,000 ft./min., or 3,000 to 4,000 ft./min. as required.

Operation. The instrument under test is mounted in a pressure chamber, the chamber being connected to the tester. A 210–240 volt, 50 c.p.s. supply is connected to the test and the main switch put to the ON position. The selector switch is set to the range required. The pressure chamber is now opened to a vacuum source via a control valve so that the rate-of-climb indicator shows an appropriate reading. Adjustment of the control valve is made until the instrument under test reads 1,000 ft./min. climb. Mercury in the manometer is now rising at a rate proportional to this rate of climb. As the mercury reaches the tip of rod 1, a contact is closed to start the synchronous clock and when the mercury reaches rod 2 a second contact stops the clock. Thus the time interval on the clock should be exactly one minute if the instrument under test is correct. If it is not correct the error is shown in seconds, and the clock dial is calibrated to show the permissible error.

Similar checks can be made at other rates of climb by use of the synchronous clock and appropriate use of the range selector switch. To check rates of descent, the pressure chamber is now closed off from the vacuum source and a second valve used to open the chamber to the atmosphere. Adjustment of this valve will allow rates of descents to be set as required.

Rate-of-climb Tester (*b*). This is a much more up-to-date tester than the one outlined under heading (*a*) and is also produced by Bryans Aeroquipment Ltd. It has already superseded the model described previously. The tester (Fig. 16.12) consists of two null-reading manometers formed by cisterns, and includes flexible tubing and glass sighting tubes. Each cistern is carried on a scale bar and can be clamped to the bar by means of a clamping knob. Provision is made for accuracy of adjustment of the cisterns by means of verniers. The scale bars are made adjustable by means of a knurled wheel which allows them to be set to the NULL position. A centre support tube on the tester carries the two glass sighting tubes in each of which is a probe connected to an electrical circuit. An electric clock is mounted into the base of the tester and is operated from the probes in the sighting tubes. Also carried on the base is a means of connecting the 210–240 volt mains supply, a switch to control the supply, and signal lamps giving a visual indication of the precise NULL points.

Principle. The two NULL indicators are connected to a common outlet and, with the suction connector open to atmospheric pressure, the cistern verniers are set to the barometric pressure of the day. The knurled scale wheels are adjusted to NULL each manometer as indicated by the signal lamps.

Now if it is required to check a rate-of-climb indicator, say, between 6,000 and 8,000 ft., the right-hand cistern is set at 6,000 ft. and the left-hand cistern at 8,000 ft. With the instrument in a vacuum chamber connected to the tester, a control panel is operated to allow the chamber to be evacuated. The evacuation rate is controlled to give a reading on the rate-of-climb indicator, say, 2,000 ft./min. Under these conditions mercury is rising in each manometer, and at 6,000 ft. the mercury in the right-hand sight glass will contact the probe to switch on the electric clock. At 8,000 ft. the mercury will have reached the probe in the left-hand manometer to switch off the clock. The time interval on the clock will show the actual rate of climb; i.e. if the clock shows 59 seconds for 2,000 ft. climb, the true rate of climb is 2,030 ft./min. approximately.

Range. The tester is supplied with reversible scale bars calibrated in the following ranges:

Front face: Left-hand edge, 0–100,000 ft. I.C.A.O. in increments of:
500 ft. up to 30,000 ft.
1,000 ft. from 30,000 ft. to 50,000 ft.
5,000 ft. from 50,000 ft. to 60,000 ft.
10,000 ft. from 60,000 ft. to 100,000 ft.

Right-hand edge, "Barometric Pressure of the Day", setting scale 31·5 in. Hg abs. to 25 in. Hg abs. A vernier on the cistern provides for precise settings to 0·002 in.

Reverse face: Left-hand edge, 0–30,000 metres I.C.A.O. in increments of:
100 metres up to 5,000 metres.
500 metres from 5,000 to 20,000 metres.
1,000 metres from 20,000 to 25,000 metres.

Right-hand edge, "Barometric Pressure of the Day", setting scale 800 mm. to 640 mm. A vernier on the cistern provides for precise settings to 0·05 mm.

Accuracy: Cistern. The bore has an extremely high finish to overcome any stiction between the cistern wall and the mercury, thus giving repeatability of readings. *Scales.* These are machine-divided to an accuracy ±0·07 mm. in 754 mm. (true millimetres).

The design of the NULL indicators is such that repeatability of readings is automatic.

Instructions for setting-up and operating the tester should be obtained from the manufacturer before commencing tests on rate-of-climb indicators.

TEST BAROMETERS

Single-cistern Barometer. A suitable standard test barometer (Fig. 16.13) is supplied by Negretti & Zambra. It consists of a single cast-iron cistern into which is inverted a glass tube with a bore of 0·5 in. Scales fitted to the barometer can show a change of pressure from 32 to 5 inches and altitude up to 60,000 ft. Alternatively, metric scales from 790 to 130 mm. and up to 18,000 metres are available. A vernier is fitted and can be controlled from a small hand-wheel at the bottom of the scale, while a thermometer indicates the temperature of the mercury in the barometer.

Connection of the barometer to a vacuum system for checking altimeters is made by means of a union which will receive a thick-walled rubber tube. A needle valve in the union can be closed to make the barometer reasonably safe during transportation. Adjusting screws are provided in the base to enable the barometer to be levelled, as indicated by two small spirit levels mounted on the base.

In Chapter 5, in which the principle of the barometer was discussed, mention was made of the various scales, standard conditions and corrections applied to the barometer. In recent years efforts have been made to reduce these various standards to one which is common throughout the world. The results of the World Meteorological Organization meetings in 1952–53 are shown in British Standard 2520 : 1954 where the new barometer conventions are clearly defined.

As an indication of the changes that have occurred, one may consider the instrumental conditions of the barometer. Under the old conditions the standard barometer readings were true only if: (*a*) the scale (assumed to be an inch scale) was at a temperature of 62 deg. F.; (*b*) the temperature of the mercury itself was 32 deg. F.; (*c*) the gravity at the measuring point was 980·62 cm./sec^2. The "London Laboratory" barometer assumed both scale and mercury to be at a temperature of 62 deg. F. and a gravitational value as at latitude 51½ deg.

British Standard 2520 : 1954 shows the new agreed convention to adopt conditions which require the whole barometer to be at 0 deg. C. and subject to a standard gravity of 980·665 cm./sec^2. New conventions for pressure relationships are shown to be:

$$1 \text{ mb.} = 0{\cdot}750062 \text{ mm. Hg} = 0{\cdot}02953 \text{ in. Hg}$$
$$1 \text{ mm. Hg} = 1{\cdot}333224 \text{ mb.} = 0{\cdot}0393701 \text{ in. Hg}$$
$$1 \text{ in. Hg} = 33{\cdot}8639 \text{ mb.} = 25{\cdot}4 \text{ mm. Hg}$$

British Standard 2520 : 1954 also provides tables of corrections to be applied for changes in temperature, latitude and altitude of the measuring point at which the barometer is situated. A study of this publication is recommended.

Twin-cistern Barometer. This is a Negretti & Zambra barometer which conforms to British Standard 2520 : 1954. It consists of a clear glass tube of 12 mm. bore diameter mounted centrally between two interconnected cisterns (Fig. 16.14). The purpose of this arrangement is to enable zero readings to be taken and obviates errors which arise from imperfect levelling of barometers having only one offset cistern. The range of the barometer enables it to be used for checking altimeters up to 100,000 ft. A vernier is supplied for use over the whole scale, with sensitive final adjustment by means of a rack-and-pinion mechanism. A white reflector gives a clear background for sighting the mercury meniscus.

Temperature corrections can be made with the aid of the thermometer mounted on the side of the barometer. Levelling of the barometer is by the same method as that used in the single-cistern type, but a circular bubble

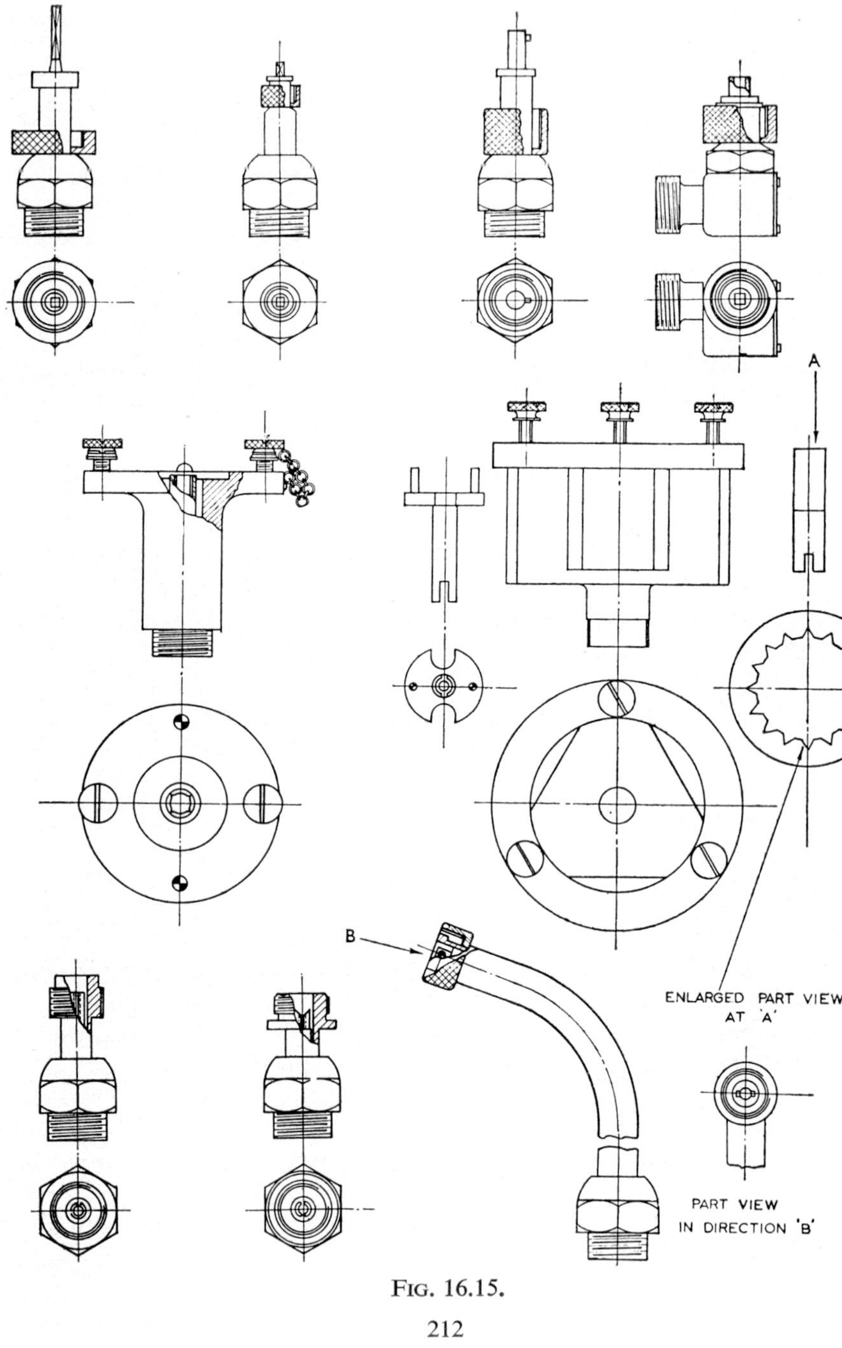

FIG. 16.15.

level is used in this instance. Connections are provided for use with $\frac{3}{16}$ in. bore rubber tubing and a branch pipe carries pressure to the top of both cisterns.

ENGINE-SPEED INDICATOR TESTING

In Chapters 8 and 13 mention was made of the methods of testing mechanical and electrical engine-speed indicators and generators. Two commercially available tachometer testers are supplied by Bryans Aero-quipment Ltd. They are (*a*) the bench-type tachometer tester and (*b*) the Master Stroboscopic Desk.

Tachometer Tester. This is a convenient tachometer tester suitable for bench mounting; it is illustrated in Fig. 13.10. The tester consists of a constant-speed motor, an infinitely variable drive and a power pack to feed a neostron lamp. Provision is made to connect a tuning fork to the tester as a means of establishing a frequency, and this provides the standard against which test-indicator readings are compared.

The indicator to be tested is connected to the tester, using a suitable adaptor. (In the case of an electrical instrument a standard generator is connected to the tester, and the indicator is connected to this.) The tester is plugged in to a 230-volt 50 c.p.s. supply and the switch at the front of the tester moved to the ON position. With the tuning fork vibrating it is possible to see a stroboscopic drum rotating behind a small window in the front of the case housing the tester. A control knob at the left of the tester enables the variable drive to be varied to change the speed indication of the connected indicator. At the same time, the stroboscope pattern changes and a chart affixed alongside the window through which the stroboscope is viewed allows the pattern to be intepreted as a fixed speed. Thus the indicator reading can be compared with this speed and errors noted. It should also be noted that the speed-control knob should not be operated unless the motor is running, and the tester should not be switched off until a low r.p.m. has been established by means of the control knob. A number of adaptors are available for connecting any type of mechanical or electrical instrument to the tester. The more common adaptors are shown in Fig. 16.15.

Master Stroboscopic Desk. Like the bench tester the Master Stroboscopic Desk (Fig. 16.16) consists of a constant-speed motor, variable drive and power pack, with a tuning fork fitted as standard equipment. In this instance, the whole arrangement forms a unit which provides also a desk at which an operator may record results of the tests.

The method of operation is similar to that of the bench tester but the stroboscope pattern is different. A disc similar to that described in Chapter 13 is used, and interpreted accordingly. It should be remembered that engine-speed indicators recording turbine speeds are geared down from the turbine.

FIG. 16.16. *Bryans Master Stroboscopic Desk in which the motor and drive are conveniently housed.*

When directly driven by the stroboscope, therefore, they will give a reading higher than that of the stroboscope by a factor dependent on the gear ratio between turbine and indicator. Suppose the turbine to have a maximum speed of 20,000 r.p.m. and a gear ratio of 4 : 1 the indicator in showing this value will, in fact, only be rotating at 5,000 r.p.m. Thus when the stroboscope is showing 5,000 r.p.m. the indicator should read 20,000 r.p.m. Similarly, when testing low-range indicators it may be necessary to divide the stroboscope reading by some factor, again dependent on gear ratio.

Testing Engine-Speed Indicator Generators. The Master Stroboscopic Desk and the bench tester can both be used for testing generators if a load test panel is available. This consists of a panel housing an a.c. voltmeter, a.c. ammeter and appropriate switches which enable varying electrical loads to be connected to the generator. The usual method of testing is to run the stroboscope at a constant speed of 3,000 r.p.m. and to measure the phase voltages with a given line current, as dictated by the load connected to the generator.

GYROSCOPE TEST TABLE

In order to test gyroscopic instruments, either electrically or pneumatically operated, it is necessary to have some kind of gyroscope test table such as the Bryans Mark 4. This provides continuous roll, pitch and yaw motion for the displacement type of gyroscope, and azimuth motion for the rate type of gyroscope. The instrument consists of a flat circular table, as shown in Fig. 9.13, on which suitable brackets are mounted for holding the instruments under test. Provision is made whereby the table can be held in the horizontal position or tilted at a fixed angle (up to $7\frac{1}{2}$ deg. maximum). A constant-speed motor drives a variable-speed unit which, in turn, drives the table. An electric clock, forming part of the test equipment, will indicate the azimuth rate of the table. The clock may also provide a control for automatic reversal of the table, once every minute, by suitable selection of a switch marked AUTO and HAND. This switch is shown at the lower left-hand side of the panel in Fig. 9.13. In the HAND position, reversal is determined manually by another switch at the lower right-hand side of the panel.

The test table is designed primarily for electrical instruments, either the 28-volt d.c. type or the 115-volt, 400 c.p.s., a.c. type. An adaptor is available so that pneumatic instruments may also be tested. This adaptor can be left in position and does not interfere with the electrical operation of the turntable.

Three adjustable handles are mounted beneath the turntable. These are not very clearly seen in Fig. 9.13, but the positions of *A* and *C* are indicated. The handles are marked *A*, *B* and *C*, and allow the table to be set for the particular operating conditions, namely:

(1) *Setting table for azimuth motion:* (i) Release control handle *A*. (ii) Lock control handle *B*. (iii) Lock table in horizontal plane.

(2) *Setting table for roll, pitch and yaw:* (i) Release handle *B*. (ii) Lock handle *A*. (iii) Release handle *C* and tilt table. (iv) Relock handle *C*.

(3) *Setting table for erection and cardinal check:* (i) Lock table in horizontal plane. (ii) Release handle *B*. (iii) Lock handle *A*.

The rate of rotation of the turntable may be varied by means of a control knob, on the left-hand side of the test table, between the limits 72 deg. to 2,880 deg. per minute.

A variety of adaptors is available to enable electrical instruments of different manufacture to be connected to the test table. A toggle switch on the front control panel (lower, centre) operates the timing mechanism over a given period of the rotation of the table to provide automatic timing for rate-of-turn indicators.

ELECTRICAL THERMOMETER TEST SETS

It is possible to test an electrical thermometer in the same manner as its mechanical counterpart; that is, with the thermometer bulb immersed in a hot-bath which contains a master mercury thermometer. This method is, however, not really practical for electrical instruments for two main reasons: (i) The process of heating the bath takes a considerable time. (ii) The system as a whole is checked by this method whereas, in fact, it is required to know whether either of the two separate parts of the thermometer system is operating effectively and within tolerance. In order to provide rapid and accurate methods of checking electrical thermometer systems, test sets have been designed specifically for this purpose.

Pyrometer Test Set. A suitable test set for checking the thermo-couple type of thermometer is the Sangamo Weston Model S137 set. This can be used to test a complete thermometer installation in the aircraft or may be used to check the separate component parts of the thermometer system; i.e. thermo-couple, extension leads and indicator. The test set is housed in a teak

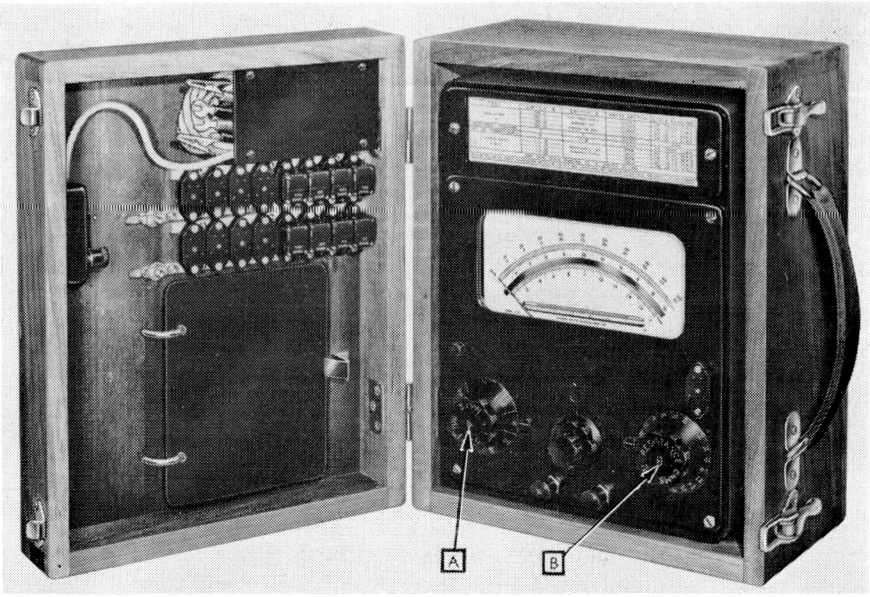

FIG. 16.17. *Sangamo Weston model S137 pyrometer test set, showing the range switches* A *and* B.

carrying case made to withstand tropical conditions. The lid of the case is detachable and provides stowage for a pair of 9 ft. leads, a unit for checking the voltage compensator of the Type B thermo-couple system, eight plug-in resistors and an instruction book. The resistors are for use in testing existing systems, but provision has been made in the case to carry additional resistors for future systems which may require different values of resistance from those supplied.

Two rotary switches *A* and *B* (Fig. 16.17) are mounted on the master instrument panel and rheostats allow fine and coarse control. Terminals allow connection of the system or parts of the system to be tested, the appropriate range selection being made on the rotary switches. Finally, the master indicator itself allows of tests to an accuracy of ±0·6 per cent or 0·06 ohm (whichever is the greater) when used for resistance testing, and ±0·6 per cent of full-scale deflection when used on the millivolt range.

When the test set is used for checking the resistance of leads, thermo-couples and so on, the rotary switch *B* is used in conjunction with the master indicator. The master indicator is directly calibrated to give a full-scale deflection for 2 ohms, while the rotary switch allows of increases of 2 ohms per step through 15 steps. Thus it is possible to measure accurately to

32 ohms by adding the value of ohms obtained on switch *B* to the indicator reading.

If it is required to test the e.m.f. of a thermo-couple, it should be connected to the test set and the millivolt output noted on the master indicator. The temperature of the thermo-couple is then compared with the millivolt reading, reference being made to the appropriate chart showing deg. C./mv, in the instruction book accompanying the test set. Care should be taken to

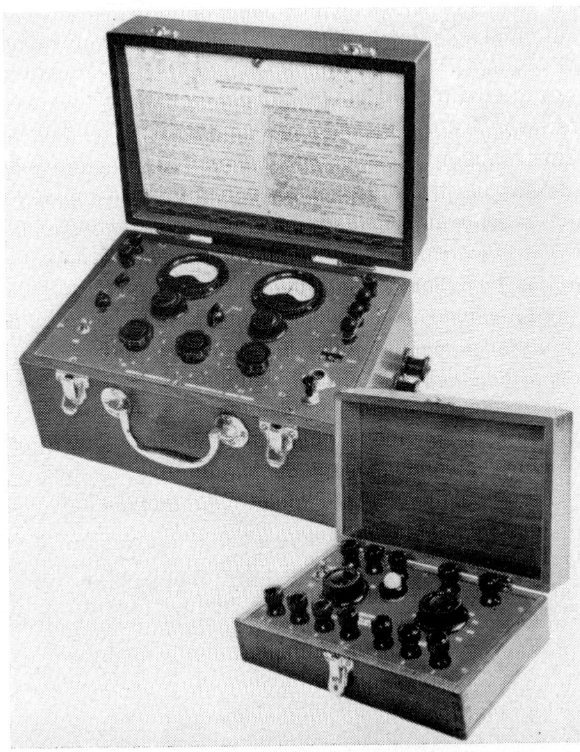

FIG. 16.18. *Bryans resistance-type thermometer test set. The adaptor for pressure gauges is shown at the right.*

correct for the ambient temperature of the test set, as indicated in the instruction book. A mercury thermometer is built into the test set for this purpose.

To test an indicator of the thermo-couple type, connect it to the terminals of the test set and select appropriate ranges on switches *A* and *B* (Fig. 16.17). Adjustments of the rheostat controls allow the readings on the indicator to be increased. Comparison is then made between the millivolt indication of the master indicator and the temperature reading of the tested instrument; again the instruction book gives the relationship between millivolts and temperature. The output of the voltage compensator on the Type B thermo-couple system may be checked by connecting it to the test set and selecting the 2-volt range on the master instrument.

In order to protect the test set should incorrect settings be made, a press-button incorporating a double-pressure device is fitted to the set. Before making any measurements, the button should be pressed lightly and the instrument deflection noted. If this is more than 10 per cent of full scale, the switch settings are incorrect; if less than 10 per cent, the required measurement may be completed by depressing the button to its fullest extent.

Ratiometer-type Test Set. Resistance thermometers of the ratiometer type require different treatment from that accorded to thermo-couple instruments because of the different principles involved. In testing this type of instrument a test set such as the Bryans No. 1407 set is required. This is housed in a teak case and is provided with a carrying handle (Fig. 16.18). The lid is detachable and is used for storing adaptors and leads used with the set. Instructions for operating the set are also contained in the lid. The test set contains a voltmeter, galvanometer, potentiometer, selector switches, terminals, two-pin and three-pin sockets, a fuse-holder and a set of decade resistors. Although the tester is designed for testing thermometers of the d.c. type it incorporates a power pack which can accept the normal 210/240-volt a.c. supply and converts this to a smoothed 30-volt d.c. supply. Alternatively, a two-pin socket on the set will accept a standard 28-volt d.c. input.

To test the ratiometer indicator, it is first connected to the test set by means of suitable adaptors. Using the appropriate table of ohms/temperature, the decade resistors are adjusted and the instrument readings noted and compared.

To test the leads and thermometer bulb, select the required adaptor so that the leads and bulb can be connected to the test set. A switch is set to "Bulb" and the d.c. supply adjusted to read 5 volts on the enclosed voltmeter. The decade resistors are adjusted until the balance position is indicated on the galvanometer, when the resistance indicated can be compared with the known resistance/temperature characteristic of the bulb.

Testing Ratiometer-type Pressure Gauges. In Chapter 12 mention was made of the fact that the ratiometer type of indicator could also be used for measuring pressure. These ratiometer pressure indicators can be tested with the ratiometer thermometer test set by making use of the Bryans Pressure Gauge Tester No. 1408, which must be used in conjunction with the Thermometer Set. Both test sets are shown in Fig. 16.18 and detailed instructions for operating either or both sets may be obtained from the manufacturer.

Jetcal Thermo-couple Tester. The Jetcal tester is available through Bryans Aeroquipment Ltd.; it is a tester which is used for checking a complete thermo-couple installation in the aircraft. This is achieved without breaking into the installation at any point and without the necessity of running the engine. The tester is used to heat the thermo-couples in the aircraft installation and to measure accurately the temperature to which they are heated. Thus, comparison with the aircraft indicator will show the inaccuracies of the particular installation. A range of temperatures up to 800 deg. C. can be obtained with an accuracy of ± 4 deg. C. at the rated engine test temperature.

The Jetcal tester is built into a metal box which can be carried out to the aircraft by means of a trolley designed for it (Fig. 16.19). An instrument panel

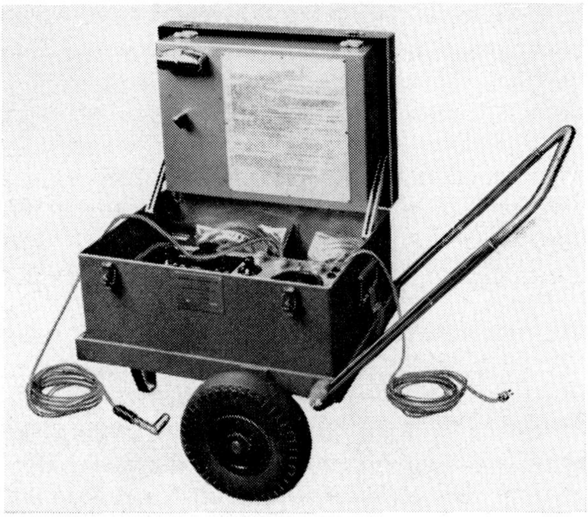

FIG. 16.19. *Jetcal thermo-couple tester with transporting trolley.*

is carried inside the box, which also provides storage space for cables. The lid of the box is used to house the thermo-couple heaters and other accessories. A variable-voltage regulator is mounted on the panel; the regulator controls the mains supply to the heaters thereby governing their temperature.

The panel also carries mains fuses, a socket for the heater leads and a change-over selector switch. This switch enables a master potentiometer, also mounted on the panel, to be used for measuring the temperature of the heaters or the temperature of the thermo-couples under test. Sockets on the panel allow cable connections to be made for calibrating cockpit indicators. There are two temperature scales on the master potentiometer, 0–500 deg. C. and 500–1,000 deg. C. Additionally, there are two millivolt scales, 0–50 mV. and

50–150 mV. The temperature scales are used in conjunction with the testing of chromel-alumel thermo-couples, while the millivolt scales can be used with thermo-couples of other materials.

Operation. A Jetcal heater is placed over each thermo-couple under test and power is applied to the heaters from the mains supply. The voltage regulator is used to adjust the temperature to that required. In the base of each of the Jetcal heaters is a thermo-couple, and the average temperature of all the heaters in use is measured with the master potentiometer. The temperature reading obtained is compared with the reading of the indicator installed in the aircraft. If any error is noted on the aircraft indicator it can be disconnected from the circuit and a cable can then be used to feed back to the master potentiometer in the Jetcal tester. Thus the master potentiometer can be used as the indicator of the aircraft installation as well as the indicator of the heater temperature. Similarly, engine-fuel systems and controls can be set accurately, with the engine running, using the master potentiometer as a standard indicator. The aircraft indicator can also be checked, using the battery and potentiometer provided, in the tester, the master potentiometer being used as a standard. The foregoing procedure can also be used when testing the installation of gas-turbine temperature controllers (see page 153).

PORTABLE PITOT/STATIC TEST SET

The test equipment described so far requires removal of the instrument from the aircraft before testing can begin. The portable pitot static test set designed by Kelvin & Hughes, however, provides a means of testing manometric instruments (i.e. airspeed indicators, machmeters and so on) *in situ* thus saving time. The test set is housed in a case containing an instrument panel on which two pressure indicators are mounted (Fig. 16.20). The panel also carries control valves, a change-over valve and a double-acting hand-operated pump to supply the pressure.

In order to provide sensitivity and accuracy of indication the two pressure instruments are servo-operated from pressure transducers; these require suitable power

FIG. 16.20. *Kelvin & Hughes portable pitot/static test set.*

Table 16.2

RANGE OF PITOT/STATIC TEST SET

Pressure System	*Indicated Pressure*	*Altitude*	*Indicated Airspeed*	*Mach No.*
Static	0·3 to 30 in. Hg (absolute)	0 to 100,000 ft.		Limiting Mach number at sea level
Pitot	0–37 in. Hg (differential)	—	80–750 knots	0–1·13

packs and amplifiers. Power for the servo loops and amplifiers comes from a 115-volt, 400 c.p.s. three-phase supply. The servo-operated indicators each carry a pointer which rotates one complete revolution for a pressure change of 1 in. Hg. Dial markings are such as to allow the instrument to be read accurately to within 0·01 in. Hg. Outlets from the test set allow for connection to an adaptor which slips over the aircraft pitot-head to provide a pressure-tight connection. The range of the equipment is shown in Table 16.2.

Operation. The adaptor is fitted over the aircraft pressure-head and connected by flexible hose to the connectors on the test set. A source of supply (115 volt, 400 c.p.s. three-phase) is connected to the test set and the hand-pump operated to provide the appropriate pressure. The pressures used on the test set depend on the particular instrument being tested and conversion tables—supplied with the set—enable these pressures to be interpreted as airspeed, Mach number and altitude. The pressure applied by the pump is fed into the aircraft pitot/static system, as determined by the change-over valve, and also to the pressure transducers. These operate on the force-balance system; i.e. the pressure-sensitive element in the transducer is held in balance by means of a spring. The application of pressure to the transducer causes an unbalance to produce an output signal from an electrical pick-off. The signal is amplified and used to drive a servo system to restore balance by changing the torque of the balancing spring. The rotation of the servo system in restoring balance is thus a measure of the pressure which produced the unbalance. A pointer moves over a dial to indicate pressure change to a sensitivity of 0·01 in. Hg.

Large-capacity Pitot/Static Systems. In the case of large-capacity systems provision is made in the test set for connecting to an external pressure source which is then used in conjunction with the hand-operated pump. When not in use the test set is protected by a lid which provides a storage space for the pressure-head adaptor and the conversion tables.

APPENDIX 1

Conversion Factors and Tables

Table 1

USEFUL CONVERSION FACTORS

Multiply	*By*	*To obtain*	*Reciprocal**
Knots	1·15152	M.P.H.	0·868
Kilometre/hour	0·62137	M.P.H.	1·60934
Kilometre/hour	0·5396	Knots	1·8532
U.S. gallons	0·83267	Imp. gallons	1·20095
Litres	0·219975	Imp. gallons	4·5459631
Kilogram/sq. cm.	14·2233	lb./sq. in.	0·07031
Inches Hg	0·4902	lb./sq. in.	2·039

* To convert the units shown in the third column to those in the first column, use the reciprocal factor.

Temperature Conversion. The two most commonly used temperature scales are those of Fahrenheit and Celsius. It should be noted that the Celsius scale is expressed as deg. C., which many people regard as meaning "deg. Centigrade". In fact, an International Commission of 1948 ruled that the term *Centigrade* (and the corresponding term *Centismal,* as used on the Continent) would no longer be used, but that all countries should use the term Celsius. In practice, the actual scale values remain the same and only the name has been changed. The usual method of conversion has already been given on page 91. Another more unusual conversion is:

$$\text{deg. C.} = \tfrac{5}{9}(\text{deg. F.} + 40) - 40$$
$$\text{deg. F.} = \tfrac{9}{5}(\text{deg. C.} + 40) - 40$$

It has the merit of having the same form whichever way the conversion is applied and is more easily remembered because of this. It is based on the fact that the two scales are identical at -40 deg. (e.g., -40 deg. F. $= -40$ deg. C.).

Table 2

PRESSURE/HEIGHT RELATIONSHIP

Height (ft.)	*Pressure (in. Hg) London Laboratory*	*Height (ft.)*	*Pressure (in. Hg) London Laboratory*
−1,000	31·09	30,000	8·91
0	29·99	31,000	8·51
+1,000	28·93	32,000	8·13
2,000	27·89	33,000	7·76
3,000	26·88	34,000	7·40
4,000	25·91	35,000	7·06
5,000	24·96	36,000	6·73
6,000	24·04	37,000	6·41
7,000	23·15	38,000	6·11
8,000	22·28	39,000	5·83
9,000	21·44	40,000	5·55
10,000	20·63	41,000	5·29
11,000	19·84	42,000	5·04
12,000	19·08	43,000	4·81
13,000	18·34	44,000	4·58
14,000	17·62	45,000	4·37
15,000	16·93	46,000	4·16
16,000	16·26	47,000	3·97
17,000	15·61	48,000	3·78
18,000	14·98	49,000	3·60
19,000	14·37	50,000	3·43
20,000	13·78	51,000	3·27
21,000	13·22	52,000	3·12
22,000	12·67	53,000	2·97
23,000	12·14	54,000	2·83
24,000	11·63	55,000	2·7
25,000	11·13	56,000	2·57
26,000	10·65	57,000	2·45
27,000	10·19	58,000	2·34
28,000	9·75	59,000	2·23
29,000	9·32	60,000	2·12

APPENDIX I

Table 3

MILLIBAR/MILLIMETRE/INCHES/HEIGHT RELATIONSHIP

Pressure (Millibars)	*Millimetres Hg*	*Inches Hg*		*Equivalent Pressure Height (ft.)*
		Standard	*London Laboratory*	
1,050	787·6	31·007	31·084	−991
1,040	780·1	30·711	30·788	−724
1,030	772·6	30·416	30·492	−455
1,020	765·1	30·121	30·195	−185
1,013·2	760·0	29·920	29·994	0
1,010	757·6	29·825	29·900	+88
1,000	750·1	29·530	29·603	363
990	742·6	29·235	29·308	640
980	735·6	28·939	29·011	919
970	727·6	28·644	28·715	1,200
960	720·1	28·349	28·419	1,484
950	712·6	28·054	28·123	1,771
940	705·1	27·759	27·827	2,060
930	697·6	27·463	27·531	2,352
920	690·1	27·168	27·235	2,646
910	682·6	26·872	26·939	2,943
900	675·1	26·577	26·643	3,242
890	667·6	26·282	26·348	3,544
880	660·1	25·978	26·052	3,849
870	652·6	25·691	25·756	4,157
860	645·1	25·396	25·460	4,467
850	637·6	25·100	25·164	4,780
840	630·1	24·805	24·868	5,097
830	622·6	24·510	24·572	5,416
820	615·1	24·215	24·276	5,739
810	607·6	23·919	23·980	6,065
800	600·1	23·624	23·683	6,393

Table 4

BOOST-PRESSURE-GAUGE UNITS

mm. Hg (absolute)	*Lb./sq. in.*	*in. Hg (absolute)*	*mm. Hg (absolute)*	*Lb./sq. in.*	*in. Hg (absolute)*
631	−2·5	24·83	1,432	+13	56·39
657	−2·0	25·85	1,484	+14	58·42
682	−1·5	26·87	1,536	+15	60·46
708	−1·0	27·89	1,587	+16	62·50
734	−0·5	28·90	1,639	+17	64·53
760	0	29·92	1,691	+18	66·57
786	+0·5	30·94	1,743	+19	68·60
812	+1·0	31·96	1,794	+20	70·64
838	+1·5	32·98	1,864	+21	72·68
863	+2·0	33·99	1,898	+22	74·71
889	+2·5	35·01	1,949	+23	76·75
915	+3·0	36·03	2,001	+24	78·78
941	+3·5	37·05	2,053	+25	80·82
967	+4·0	38·06	2,105	+26	82·86
993	+4·5	39·08	2,156	+27	84·89
1,019	+5·0	40·10	2,208	+28	86·93
1,044	+5·5	41·12	2,260	+29	88·96
1,070	+6·0	42·14	2,311	+30	91·00
1,096	+6·5	43·15	2,363	+31	93·04
1,122	+7·0	44·17	2,415	+32	95·07
1,148	+7·5	45·19	2,467	+33	97·11
1,174	+8·0	46·21	2,518	+34	99·14
1,200	+8·5	47·23	2,570	+35	101·18
1,225	+9·0	48·24	2,622	+36	103·22
1,251	+9·5	49·26	2,673	+37	105·25
1,277	+10·0	50·28	2,725	+38	107·29
1,329	+11·0	52·32	2,777	+39	109·32
1,381	+12·0	54·35	2,829	+40	111·36

Table 5

CALIBRATION OF MACHMETERS (STANDARD CONDITIONS)

Altitude		*0 ft.*		*10,000 ft.*		*20,000 ft.*		*30,000 ft.*		*40,000 ft.*		*50,000 ft.*	
Mach No.	$\frac{(P-S)^*}{S}$	*in. Hg*	*mm.* H_2O	*in. Hg*	*mm.* H_2O	*in. Hg*	*mm.* H_2O	*in. Hg*	*mm.* H_2O	*in. Hg*	*mm.* H_2O	*in. Hg*	*mm.* H_2O
		29·921	10,341·5	20·577	7,112·0	13·74	4,752·5	8·886	3,071·0	5·538	1,914·0	3·426	1,184·0
0·7	0·3871	11·582	4,003·0	7·965	2,753·0	5·323	1,839·5	3·440	1,189·0	2·144	741·0	1·326	458·5
0·8	0·5243	15·687	5,422·0	10·788	3,729·0	7·210	2,491·5	4·659	1,610·5	2·904	1,003·5	1·796	621·0
0·9	0·6913	20·684	7,149·0	14·225	4,916·5	9·507	3,285·5	6·143	2,123·0	3·828	1,323·5	2·368	818·5
1·0	0·8929	26·714	9,234·0	18·373	6,350·5	12·278	4,243·5	7·934	2,742·5	4·945	1,709·0	3·059	1,057·5
1·1	1·1328	33·891	11,715·0	23·310	8,056·5	15·576	5,383·5	10·066	3,479·0	6·273	2,168·5	3·881	1,341·5
1·2	1·4075	—	—	28·962	10,010·0	19·353	6,688·0	12·506	4,323·0	7·795	2,694·0	4·822	1,666·5
1·3	1·7136	—	—	—	—	23·562	8,143·5	15·226	5,263·0	9·490	3,280·0	5·870	2,029·0

* Where $(P - S)$ is the dynamic pressure, and S is the static pressure.

Example. With static pressure equal to the standard atmosphere (29·921 in. Hg) and dynamic pressure 11·582 in. Hg: $(P - S)/S = 11{\cdot}582/29{\cdot}921 = 0{\cdot}3871$. This corresponds to a machmeter reading of 0·7.

Table 6

AIRSPEED/PRESSURE EQUIVALENTS

For instruments calibrated to formulae:

$$P = 0{\cdot}012504 V^2{}_1 (1 + 0{\cdot}43 V^2{}_1 \times 10^{-6})$$
$$\text{or } P = 0{\cdot}016580 V^2{}_2 (1 + 0{\cdot}57 V^2{}_2 \times 10^{-6})$$
where P = mm. of water; V_1 = m.p.h. and V_2 = knots

Airspeed V_1 *or* V_2	*mm.* H_2O V_1	*mm.* H_2O V_2	*Airspeed* V_1 *or* V_2	*mm.* H_2O V_1	*mm.* H_2O V_2
20	—	6·63	240	738·0	986·3
30	11·26	14·93	250	802·5	1,073·0
40	20·00	26·55	260	869·8	1,164·0
50	31·29	41·51	270	940·1	1,259·0
60	45·08	59·81	280	1,013·0	1,358·0
70	61·40	81·47	290	1,090·0	1,461·0
80	80·24	106·50	300	1,169·0	1,569·0
90	101·60	134·90	320	1,337·0	1,797·0
100	125·60	166·70	340	1,517·0	2,043·0
110	152·10	202·00	360	1,711·0	2,308·0
120	181·20	240·70	380	1,918·0	2,591·0
130	212·80	282·90	400	1,238·0	2,895·0
140	247·10	328·60	420	2,373·0	3,219·0
150	284·00	377·80	440	2,622·0	3,564·0
160	323·60	430·60	460	2,887·0	3,931·0
170	365·80	487·00	480	3,166·0	4,322·0
180	410·70	547·10	500	3,462·0	4,735·7
190	458·40	610·80	520	3,774·0	5,174·1
200	508·70	678·30	540	4,103·0	5,638·2
210	561·90	749·60	560	4,450·0	6,129·2
220	617·80	824·60	580	4,815·0	6,646·7
230	676·50	903·50	600	5,198·3	7,193·6

Note. The formula used to calculate the values given in Table 6 is obtained from an approximation. At low airspeeds the approximate formula is sufficiently accurate, but the accuracy falls off with increasing airspeed. Thus for high-range airspeed indicators a full adiabatic formula is used:

$$P_i = P_o \left[\left(1 + \frac{\gamma - 1}{2\gamma} \cdot \frac{e_o}{P_o} \cdot V_r^2 \right)^{\frac{\gamma}{\gamma - 1}} - 1 \right]$$

where P_o is the impact pressure, P_o is the static pressure, γ is the ratio specific heat constant pressure/constant volume, e_o is the air density (standard sea-level conditions), and V_r is the rectified airspeed. The values for the Pressure/Velocity Table then become as shown above.

APPENDIX I

Table 7: PRESSURE/VELOCITY TABLE

Airspeed (Knots)	*mm.* H_2O	*Airspeed (Knots)*	*mm.* H_2O	*Airspeed (Knots)*	*mm.* H_2O
40	26·55	240	986	420	3,231
80	106·40	250	1,073	440	3,580
90	134·90	260	1,165	460	3,953
100	166·70	270	1,260	480	4,349
110	201·90	280	1,359	500	4,771
120	240·60	290	1,462	520	5,219
130	282·80	300	1,570	540	5,693
140	328·50	310	1,683	560	6,196
150	377·70	320	1,799	580	6,734
160	430·60	330	1,921	600	7,301
170	487·00	340	2,046	620	7,897
180	546·80	350	2,177	640	8,532
190	610·80	360	2,312	660	9,202
200	678·30	370	2,452	680	9,922
210	749·60	380	2,598	700	10,663
220	824·90	390	2,748	720	11,446
230	903·60	400	2,904	740	12,259

Table 8: TEMPERATURE/RESISTANCE CHARACTERISTICS FOR AMERICAN NO. 1 (EDISON) ELEMENTS IN RESISTANCE THERMOMETERS

t	*R*	*t*	*R*
−70	68·27	+120	137·78
−60	71·19	+130	142·40
−50	74·24	+140	147·11
−40	77·39	+150	151·91
−30	80·56	+160	156·79
−20	83·77	+170	161·89
−10	87·04	+180	167·13
0	90·38	+190	172·46
+10	93·80	+200	177·95
+20	97·31	+210	183·55
+30	100·91	+220	189·40
+40	104·60	+230	195·35
+50	108·39	+240	201·55
+60	112·28	+250	208·00
+70	116·27	+260	214·60
+80	120·36	+270	221·50
+90	124·55	+280	228·45
+100	128·85	+290	235·55
+110	133·26	+300	242·70

Temperature, *t*, is given in deg. C. Resistance, *R*, is in ohms.

Table 9: TEMPERATURE/RESISTANCE CHARACTERISTICS FOR NICKEL ELEMENTS IN RESISTANCE THERMOMETERS

t	*R*	*t*	*R*
−80	58·80	+40	108·48
−70	62·28	+50	113·40
−60	65·88	+60	118·44
−50	69·60	+70	123·60
−40	73·44	+80	128·88
−30	77·40	+90	134·28
−20	81·48	+100	139·80
−10	85·68	+110	145·44
0	90·00	+120	151·20
+10	94·44	+130	157·08
+20	99·00	+140	163·08
+30	103·68	+150	169·2
		+160	175·44

Temperature, *t*, is given in deg. C. Resistance, *R*, is in ohms.

Table 10: TEMPERATURE/RESISTANCE CHARACTERISTICS FOR PLATINUM ELEMENTS IN RESISTANCE THERMOMETERS

t	*R*	*t*	*R*	*t*	*R*
−100	77·67	+100	180·70	+300	277·55
−90	83·00	+110	185·69	+310	282·24
−80	88·30	+120	190·66	+320	286·90
−70	93·58	+130	195·61	+330	291·56
−60	98·83	+140	200·56	+340	296·20
−50	104·07	+150	205·48	+350	300·82
−40	109·29	+160	210·39	+360	305·43
−30	114·49	+170	215·29	+370	310·02
−20	119·68	+180	220·17	+380	314·60
−10	124·85	+190	225·03	+390	319·16
0	130·00	+200	229·88	+400	323·71
+10	135·14	+210	234·72	+410	328·24
+20	140·26	+220	239·54	+420	332·75
+30	145·37	+230	244·34	+430	337·26
+40	150·46	+240	249·13	+440	341·74
+50	155·54	+250	253·91	+450	346·21
+60	160·60	+260	258·67	+460	350·67
+70	165·65	+270	263·41	+470	355·11
+80	170·68	+280	268·14	+480	359·54
+90	175·70	+290	272·85	+490	363·95

Temperature, *t*, is given in deg. C. Resistance, *R*, is in ohms.

APPENDIX II

British Standard G. 100

GENERAL REQUIREMENTS FOR ELECTRICAL EQUIPMENT AND INDICATING INSTRUMENTS FOR AIRCRAFT

THIS British Standard* is divided into two parts. Part I is concerned with general requirements for equipment, including indicating instruments. Part II contains additional requirements for indicating instruments.

There are four sections to Part I as follows:

Section 1. Requirements relating to operating conditions.
Section 2. Materials and processes.
Section 3. Particular design requirements.
Section 4. Tests.

Part II deals with the construction of indicators and their cases and connections to the cases. Chapters are also included on dials, pointers, dial markings, cover glasses, sealing rings and manufacturers' identification marks. The Standard contains a number of appendices on:

(*a*) Volts-drop in cables.
(*b*) Recommended practices for the avoidance of corrosion.
(*c*) Test for flameproof equipment.
(*d*) Tests for fire-resistance and low inflammability.
(*e*) Measurement of compass safe-distance.
(*f*) Impact acceleration tests.

Graphs and illustrations of various tests are included, and tables are provided giving operating conditions of temperature, acceleration, insulation, vibration fatigue and resonance tests. A number of these tables are given here for reference. As B.S. G.100 is, however, under review at the time of publication of this book, none of the figures quoted should be used without making further reference to the British Standards Institution.

* Reference to B.S. G.100 is made by permission of the British Standards Institution, 2 Park Street, London, W.1, from whom copies of the complete Standard may be obtained, price 10s. (Please see note on page 236.)

MINIMUM VALUES OF INSULATION RESISTANCE

Type of equipment	Insulation resistance (minimum)			
	Between insulated line parts		To earth	
	Standard tests (Clause 24)	Climatic tests (Clause 25)	Standard tests (Clause 24)	Climatic tests (Clause 25)
A. Wiring components such as terminal blocks, connecting plugs, sockets and switches	100 megohms	100 megohms	$\frac{\text{200 megohms}}{\text{No. of terminals connected together}}$	$\frac{\text{200 megohms}}{\text{No. of terminals connected together}}$
B. Rotating machinery (excluding indicating instrument equipment)	20 megohms	0·5 megohm or $\frac{\text{rated voltage} \times \text{megohms}}{150}$ whichever is the greater	20 megohms	0·5 megohm or $\frac{\text{rated voltage} \times \text{megohms}}{150}$ whichever is the greater
C. Indicating instruments and equipment other than Types A and B	20 megohms	5 megohms	20 megohms	5 megohms

(*a*) The minimum insulation resistance between insulated live parts shall be the minimum insulation resistance between any pair of adjacent conducting paths insulated from one another. (*b*) Equipment with an insulated mounting shall be tested on a metal plate.

RESONANCE TESTS

Phase	Test Instruction	*Grade 1*		*Grade 2*		*Grade 3*	
		Vertical and lateral	*Longitudinal*	*Vertical and lateral*	*Longitudinal*	*Vertical and lateral*	*Longitudinal*
A	The frequency shall be varied from 10 to 30 c/s with the vibration table set to give an amplitude of:	0·014	0·008	0·012	0·010	0·008	0·004
B	The frequency shall be varied from 30 to 80 c/s with the vibration table set to give an amplitude of:	0·008	0·005	0·006	0·005	0·005	0·003
C	The frequency shall be varied from 80 to 150 c/s with the vibration table set to give an amplitude of:	0·004	0·003	0·003	0·002	0·003	0·002
D	The frequency shall be varied from 150 to 300 c/s with the vibration table set to give an amplitude of:	0·002	0·001	—	—	—	—
E	The frequency shall be varied from 300 to 500 c/s with the vibration table set to give an amplitude of:	0·001	0·001	—	—	—	—

SHORT-DURATION AND ENDURANCE TESTS, FREQUENCIES AND AMPLITUDE

Vibration grade	*Amplitude in inches at stated frequency and direction*							
	20 c/s		*50 c/s*		*150 c/s*		*500 c/s*	
	Vertical and lateral	*Longitudinal*	*Vertical and lateral*	*Longitudinal*	*Vertical and lateral*	*Longitudinal*	*Vertical and lateral*	*Longitudinal*
1	0·015	0·008	0·011	0·007	0·003	0·003	0·001	0·001
2	0·015	0·011	0·009	0·007	0·003	0·002	0·001	—
3	0·008	0·005	0·006	0·004	0·002	0·002	0·001	0·001

Amplitude is expressed as half the total displacement.

DURATION AND FREQUENCY OF FATIGUE TESTS

Grade	*Frequency c/s* $\pm$*10 per cent*	*Total duration (hours)*
1	20 50 500	20 30 10
2 and 3	20 50 150	20 30 10

EMERGENCY ALIGHTING CONDITIONS

Upwards	5g
Downwards	7g
Sideways	$3\frac{1}{2}$g
Backwards	14g
Forwards	14g

APPENDIX II

ACCELERATION GRADES

Acceleration grade	1	2	3
Upwards acceleration limit	8g	5g	$3\frac{1}{2}$g
Downwards acceleration limit	4g	2g	$1\frac{1}{2}$g
Horizontal acceleration limit	3g	2g	2g

TEMPERATURE CONDITIONS

Temperature	*Mounting conditions*	*Test requirements*
a. +125 deg. C.	On engines where the installation is closely cowled and under extreme conditions.	The equipment shall not be deranged or damaged when subjected to this temperature for the equivalent of a period of 10 hours under full working conditions, and shall withstand temperature changes of 10 deg. per minute up to the maximum temperature.
b. +110 deg. C.	In the engine bay, equipment not blast-cooled and under overload conditions.	The equipment shall not be permanently deranged or damaged when subjected to this temperature for a period of 10 minutes under full working conditions and shall withstand temperature changes of 10 deg. per minute up to this maximum temperature.
c. +100 deg. C.	In the engine bay, equipment not blast-cooled and under normal flight conditions.	All equipment shall function normally when subjected to this temperature for the equivalent of a period of 10 hours under full working conditions followed by the conditions of (b) above, but permissible errors or deviations from standard performance will in general be greater than those allowed at room temperature (10 to 20 deg. C.).
d. +70 deg. C.	Under extreme climatic conditions on the ground.	No equipment shall be permanently deranged or damaged when subjected to this temperature for the equivalent of a period of 10 hours.
e. +50 deg. C. −20 deg. C.	In the airframe under normal flight conditions.	All equipment shall function normally over this range. Permissible errors or deviations from standard performance will in general be greater than those allowed at room temperature (10 to 20 deg. C.).

TEMPERATURE CONDITIONS (*Continued*)

Temperature	*Mounting conditions*	*Test requirements*
f. −40 deg. C.	Extreme climatic conditions on the ground.	No equipment shall be permanently deranged or damaged when subjected to this temperature for the equivalent of a period of 10 hours.
g. −70 deg. C.	In exposed positions on the aircraft.	(i) The equipment shall not be permanently deranged or damaged when subjected to this temperature for the equivalent of a period of 10 hours under full working conditions. (ii) Where equipment is composed of component parts, some of which may be in exposed positions (e.g. the capillary of a transmitting pressure gauge or thermometer) the permissible errors specified for the indicator shall not be exceeded between −20 deg. C. and +50 deg. C., nor when the exposed part is subjected to an ambient change of 10 deg. per minute within the range −70 to +50 deg. C., nor at any temperature within this range.

VOLTAGE DROP IN RUBBER CABLES, AND THERMOPLASTIC CABLES OF THE "VIN" AND "CEL" TYPES

Nominal current rating (Amp.)	*One-minute-ratings*					
	4-7 cables		*2-3 cables*		*Single cable*	
	Amp.	*Voltage drop*	*Amp.*	*Voltage drop*	*Amp.*	*Voltage drop*
2·5	9	9·2	—	—	—	—
4	16	16·3	—	—	—	—
7	25	14·6	—	—	—	—
19	52	11·0	—	—	—	—
37	135	9·3	140	9·7	—	—
64	290	7·4	305	7·8	305	7·8
83	400	6·8	450	7·7	455	7·7
138	650	6·4	745	7·3	750	7·3

APPENDIX II

VOLTAGE DROP IN RUBBER CABLES, AND THERMOPLASTIC CABLES OF THE "VIN" AND "CEL" TYPES

Nominal current rating (Amp.)	*Five-minute ratings*					
	4-7 cables		*2-3 cables*		*Single cable*	
	Amp.	*Voltage drop*	*Amp.*	*Voltage drop*	*Amp.*	*Voltage drop*
2·5	5·5	9·2	—	—	—	—
4	9	9·2	—	—	—	—
7	12	7·0	—	—	—	—
19	25	5·3	—	—	—	—
37	64	4·4	69	4·8	—	—
64	155	4·0	170	4·3	170	4·3
83	220	3·7	230	3·9	230	3·9
138	360	3·5	390	3·8	400	3·9

VOLTAGE DROP IN RUBBER CABLES, AND THERMOPLASTIC CABLES OF THE "VIN" AND "CEL" TYPES

Nominal current (Amp.)	*Continuous ratings*					
	4-7 cables		*2-3 cables*		*Single cable*	
	Amp.	*Voltage drop*	*Amp.*	*Voltage drop*	*Amp.*	*Voltage drop*
2·5	5	8·4	—	—	—	—
4	7	7·1	—	—	—	—
7	9	5·2	—	—	—	—
19	18	3·8	—	—	—	—
37	35	2·4	46	3·2	—	—
64	64	1·6	85	2·2	110	2·8
83	80	1·4	105	1·8	130	2·2
138	125	1·2	150	1·5	215	2·1

Voltage drop is calculated in volts per 100 ft. of conductor when carrying the current indicated in the column preceding. The conductor is assumed to be at 70 deg. C.

Note.—It is anticipated that one of the main differences between the current B.S. G.100 and the forthcoming edition will be an increase in the range of temperatures covered and which instruments of certain types will have to withstand. Also, in view of the widely differing weather conditions encountered by modern aircraft, it is probable that new icing and weather-proofing tests will be instituted.

Index